Nanostructure Science and Technology

Series Editor:
David J. Lockwood, FRSC
National Research Council of Canada
Ottawa, Ontario, Canada

More information about this series at http://www.springer.com/series/6331

Yi Ge • Songjun Li • Shenqi Wang
Richard Moore

Editors

Nanomedicine

Perspectives of Nanomedicine

Volume 2

 Springer

Editors
Yi Ge
Centre for Biomedical Engineering
Cranfield University
Bedfordshire, United Kingdom

Songjun Li
School of Materials Science and Engineering
Jiangsu University
Zhenjiang, China

Shenqi Wang
Advanced Biomaterials and Tissue
    Engineering Centre
College of Life Science and Technology
Huanzhong University of Science
    and Technology
Wuhan, China

Richard Moore
Biomimesis
Melton Mowbray
Leicestershire
United Kingdom

ISSN 1571-5744   ISSN 2197-7976 (electronic)
ISBN 978-1-4614-2139-9   ISBN 978-1-4614-2140-5 (eBook)
DOI 10.1007/978-1-4614-2140-5
Springer New York Heidelberg Dordrecht London

Library of Congress Control Number: 2014954585

Printed on acid-free paper

Springer is part of Springer Science+Business Media (www.springer.com)

Contents

Chapter 13
Nanomedicine: A Hyper-expectation and Dawning Realisation?

Ferdia Bates

13.1 Introduction

At this point in time, it could be assumed that to some extent nanotechnology, defined as the manipulation of material and the development of structures at a scale of between 1 and 100 nm, has to some extent permeated the lives of the all within the first world through at least one of its numerous applications. Perhaps the most poignant of these applications is that within the medical field where nanomedicine affects a person's most intimate possession, their health. This may be in something as seemingly mundane as the titanium oxide nanoparticles contained within sunscreens or as striking as increasingly effective cancer treatments, prodrugs and nanopharmaceutics. Whatever the impression that nanomedicine leaves, it is unequivocally present within society with increasing levels of sophistication.

The degree to which we rely on nanotechnology already or expect to rely on it in the future is a moot point. For years we have been conditioned to expect enormous things from science and technology, and thus far this expectation has been satisfied. Technological advancement has allowed a single generation to witness drastic changes in society the likes of which have never before been seen. Within the twentieth century, the fundamental ideas and values of this global society have undergone a radical upheaval which has occurred in hand with technological advancement. The perception of fantastical and seemingly endless innovation and technological growth is due in part to the human trait of linear thinking. Isaac Newton once demonstrated this, and his modesty, in the highly quotable line:

If I have seen further it is only by standing on the shoulders of giants

F. Bates (✉)
Universitat Autònoma de Barcelona, Barcelona, Spain
e-mail: ferdia.bates@mariecurie.cat

© Springer Science+Business Media New York 2014
Y. Ge et al. (eds.), *Nanomedicine*, Nanostructure Science and Technology,
DOI 10.1007/978-1-4614-2140-5_13

This single sentence can be taken as one of the most accurate paradigms for human thought in the sense that a mind can only imagine future technology based on the extrapolation of the state-of-the-art in the present thus the analogy of seeing ahead from a static, albeit lofty, point. The idea of fantastical technological advancement depicted by a current state-of-the-art can be demonstrated very effectively by the evolution of the popular conception of robotics with respect to time; this can be seen in the advancement of technology projected in yesteryear and the actual realisation of robotics in the present day. These predictions now seem comical to this current generation who can see the juxtaposition of the prediction beside the advancement actually achieved in the same time period (Fig. 13.1). In Fig. 13.1, the right depicts two modern examples of function and convenience while the left shows the now-anachronistic 'future' technology where the humanoid robot is set against a back drop of a flying-saucer spaceship on some non-disclosed foreign planet. The automaton in image is inspired by 'Robby the Robot', star of the films 'Forbidden Planet' (1956) and 'The Invisible Boy' (1957), who is perhaps one of the most recognisable examples of what a robot was thought to be in the 1950s, along with B9 from the series 'Lost in Space' (1965–1968), both of which – or 'whom' – were designed by the renowned Robert Kinoshita. These robots displayed personalities and wit with a level of cognition that still soars above current computing power [1, 2]. As the function of these celebrity robots was, at the core, to help and convenience their human masters, one could argue that this modern age has indeed, to some

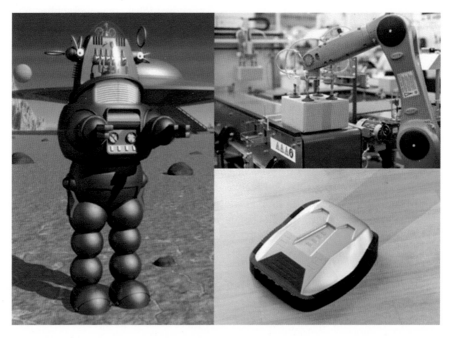

Fig. 13.1 The contrast between past predictions and present realisations of robotics technology (Copyright free images)

extent, seen the realisation of such predictions and yet there is the contrasting reality that one would still not expect anything close to the comic cornucopia displayed by those celebrity-bots from a modern day robotic vacuum.

The consolation for a conversationally barren non-humanoid vacuum cleaner is the gargantuan supply of 'unforeseen' novel technology and innovation which these predictions didn't even take account of such as personal computers, the internet, laptops, mobile phones and smart phones – a marriage of all the aforementioned technologies and something that has affected society in an unprecedented manner.

This is a good example of a paradigm shift idea, the manner in which the future is thought to be as given by the projections of the time in contrast to how that future actually transpires to be when such a time arrives with all the novel innovations it brings with it. In this way, though hyper-expectation is often less of a matter of over-exaggeration and more of misdirection, dawning realisation can be expected to bring realisation of new unforeseen technologies as well as the cul-de-sacs of optimistic speculations and attenuators associated with implementation of an ideal. This trait is ingrained in the collective mind of each generation and is the ultimate dictation as to how one can perceive the future; it is a trait compounded by the increasingly large generational gaps which technological progress has brought.

The acceleration of technological advancement with respect to time can be identified as far back as the industrial revolution, though progress as striking as was seen in the twentieth century was not realised until the advent of the computer and the massive increment in computational capacity it facilitated from its predecessor and the new technologies that were built from it; first with the use of thermionic valves, which gave rise to widespread domestic radio as well as the first electronic digital computers, and then with transistor-based modern computing. The transistor, more so than even the thermionic valve which immediately preceded it, allowed for the accelerative progress which is now taken as normality.

This accelerative progress has been embodied in Moore's law which states that computing power will approximately double every 18 months [3]. Needless to say, extrapolation of an exponential function quickly leads to quite dizzying heights relative to a static position. With such increasing computing power, technological intricacy has exploded at an extraordinary rate; an example of this is achieving space flight less just a few short decades after the achieving flight itself. Thus humanity has grown accustomed to such fantastical changes within a lifetime to the extent that it is now expected.

The advent of nanotechnology has renewed this expectation if not increased it even more. It is true to say that this age is truly one of the futurists; never before has a generation looked forward in time with such a degree of expectation. As a generation we have become accustomed to jumping from the shoulders of giants rather than merely standing on them. Though this ethos is extremely useful and the basis of this modern and highly progressive society, the level to which the culture of hyper-expectation should be entertained has always been a hot topic.

Nanomedicine has proven its worth already in its various applications and the projections for its growth and development predict sophistication and effectiveness that will radically change the way medicine is practiced [4–7]. The question that immediately follows such predictions almost always relates to the time scale involved. With the initial assimilation of nanomedicine into society approaching completion, to some degree, the mystery that had shrouded nanomedicine in the past has lifted to leave a clearheaded realisation of the work that is required before any of the grand promises that nanomedicine has given can be realised.

This being said, the economic implications of first and second generation (Fig. 13.2) nanotechnology-incorporating products can already be seen in the market place. Heavy financial investment has been made in increasingly large instalments on the back of the projected treatments and technologies that nanomedicine will yield. The National Nanotechnology Initiative in the US, having been set up initially in 1998, was designed to encourage collaboration between the fields of science, healthcare, engineering and technology on the nanoscale [9]. As of 2008, the budget requested was $1.5 billion, almost triple that spent in 2001. This figure has been revised to $1.8 billion for the 2013 in a budget proposal put forward by President Obama. This aggressive growth in the funding is indicative of the volume of research that has taken place in that short period of time [10]. Regarding nanomedicine specifically, substantial research initiatives have been set up such as that in the American National Cancer Institute's $144 million cancer nanotechnology initiative, which commenced in 2004, in order to bolster the interest in nano-research. In 2010 this initiative entered into its second phase in which many of the researched nanomedicines would be brought to clinical trial fortified by funding of equal rigor [11].

Indeed, these sums of money were offered in such quantities because of the projected growth and potential that the nanotechnology sector was speculated to have; it was predicted in 2000 that nanotechnology-incorporating devices would reach the one trillion (10^{12}) dollar mark by 2015 at which point it would account for more than two million high skill jobs [8]. In contrast to this, analysts in 2004 projected a

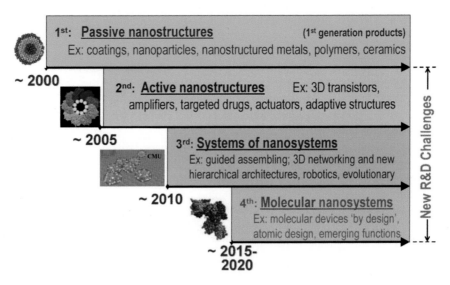

Fig. 13.2 The projected progression of Nanotechnology from 2000 to 2020 (Reproduced from [8])

nanotechnology market worth of 2.6 trillion dollars with ten million associated manufacturing jobs by 2014 [12]. This projection was based on an estimate of the 2004 market value of nanotechnology which was cited as 158 billion dollars, 92 % of which were established materials with nanoscale dimensions. Based on this valuation, a further eight fold increase in market value was projected in the decade following the publication of the report. Within this extrapolated valuation, it was estimated that 89 % of the future revenue of the market would come from new and emerging nanotechnology [13]. As of 2009, nanotechnology was estimated to be worth one quarter of a trillion dollars and, with retrospective analysis, showing a doubling of the market every 3 years as a result of the on-going introduction of new nanotechnology-incorporating products [14]. Such a revision in estimation demonstrates the highly dynamic and fast paced nature of the nanotechnology sector. Though, as demonstrated in the 2004 analysis, a large amount of the initial growth came from an increase in market share through merely nano-sizing existing components rather than through true nanotechnological innovation. Thus, it is often an arduous task to differentiate between true advancement in the field and mere colonisation of an existing market.

Having achieved increasingly complex structures in keeping with the projected timetable, prototypically, the molecular transistor which was predicted circa 2010 [15], predictions for what future nanomedicine will yield can be divided in two. The first is the conceptual research which is discovering the extraordinary applications and capabilities that nanomedicine has to offer. Overlaying this is the stark reality of attempting to translate the complex laboratory ideal of medicine on a nanoscopic scale from concept to clinic.

The nano-age offers a unique opportunity to stand starry eyed idealism, stemming from on-going conceptual research of a young field, next to the cold logistics of implementation through which is discovered the new challenges that nanomedicine will bring which can be represented by the nanotoxicity example (Sect. 13.5). The combination of these two outlooks has created a fascinating cocktail of on-going fundamental research in the face of the emergence of nanotoxicology as the main bottleneck for implementation. The rapidity of the field has also challenged the traditionally conservative and slow-moving gears of bureaucracy in determining the manner in which such an unknown and emerging sector should be governed and regulated.

Thus there is a complex ballet being performed between the hyper-expectation of the applications of nanomedicine and the dawning realisation of what the act of applying it actually entails.

13.2 Popular Media: Feeding Frenzy

There has long been a complex relationship between the popular media and science with both sides looking to the other for inspiration. Such a complexity frequently leads to confusion as to what is science fiction and what is science fact. Through the ages, this has led, to some extent, to a convoluted public perception of the state-of-the-art. On the other hand, the media often borrows from real scientific phenomena and in doing so, serves to acclimatise the public to such concepts as well as advertising the technology; in this respect, science and the media can be likened to feuding siblings pitted in a perpetual game of chase.

For many years, nanotechnology has provided fertile ground for science fiction to explain their farfetched devices and techniques. The foundations on which nano-medicine are based on can be traced even further back than this. One of the first popular media acknowledgements of a core nanomedicine concept can be seen in the film 'Dr Ehrlich's magic bullet', released in 1940, it documents the arduous journey of Dr Paul Ehrlich, the Nobel laureate often referred to as the father of modern chemotherapy and widely recognised as to be the first to propose the con-cept of targeted delivery, a central application of nanomedicine [16], to develop the first chemotherapeutic. Further homage, albeit unwittingly, to the concepts of nano-medicine can be seen in the film 'Fantastic Voyage', released in 1966, it came just 7 years after Richard Feynman's landmark talk 'There's plenty of room at the bottom' which can be taken as the conceptual birth of nanotechnology [17]. The Fantastic Voyage, perhaps aided by the pleasant visage of Raquel Welsh, has become one of the most enduring epitomes of the concepts being realised through nanomedicine. The portrayal given by this film can be greatly likened to the concept of the nanofac-tory which is already under development today (Sect. 13.3.3).

The 'Grey Goo' scenario was first proposed by nanotechnological visionary Eric Drexler in 1986 [18] as a hypothetical scenario whereby he observed the conse-quence of creating an organism that could proliferate uncontrollably, eventually engulfing all things in 'Grey Goo'. Indeed, the ability of an object to self-replicate was cited as one of the primary factors to make nanotechnologies 'particularly con-sequential' [19]. This projection built on Feynman's original proposal made almost 30 years previously. The grey goo scenario is a prime example of hyper-expectation of a conceptual field. 1986, close to 30 years ago, was still a long time short of any meaningful realisation of the concepts dealt with by Drexler. Without this 'real world' perspective, the then largely unknown field of nanotechnology could be viewed as the herald for the imminent arrival of Huxley's 1932 novel 'Brave New World' [20] bringing with it some understandable misgivings if not hysteria.

In more recent times, nanotechnology has even acquired its own 'Jaws'. This is Michael Crichton's superbly written 'Prey' [21] in which the more ominous poten-tial of sophisticated nanotechnology is explored. Nanotechnology is also very much present in modern media as well where it is used as a one word explanation of highly evolved and sophisticated technology. It played a pivotal role in the descrip-tion of many aspects of the 2009 film 'G.I. Joe' and also can be identified through-out modern popular media as the facilitator of the future advancement of medicine. The videogame 'Deus EX', the latest addition to the series having been released in 2011, depicts the central character as being an 'augmented human', heavily modi-fied with bionic appendages all of which contain nanotechnology. There is also reference made to so-called 'nanites' which are nano-robots, not unlike those specu-lated to perpetuate the 'grey goo' scenario, these robots are introduced into the blood stream of the central character where they serve to repair injuries as they occur [22].

The content of Deus EX, which is cited purely as a prototypical example, can be used as an illustration of the realisation, which has come predominantly through media-facilitated acclimatisation, and corresponding acceptance of the public of nanotechnology as a non-malevolent 'part of the woodwork' of the future.

13.3 The Dawning Implication of Hyper-expectation

What is medicine? A basic definition can be given as the maintenance of the physical body with a primary endpoint being staving off or guarding against death. The next barrier after the conquering of illness is the defeat of death itself or, in a more immediate and plausible way, to keep the reaper at bay for increasingly an increasingly long period of time. Even using the most prudent mind-set, with the medical advances which are occurring in this age, it is not unreasonable to predict a time in the not-entirely-distant future when death, certainly for those of means at least, will be optional.

Given the current trends in population age (Fig. 13.3), it is quite reasonable to state that the coming generation, as a consequence of the advances made within nanomedicine today, will be the first with the true ability to 'outlive' the very bodies which the personality of the person is contained. Indeed, the implications of nanomedicine stretch as far as one would care to look. This is already apparent in the increased prevalence of chronic degenerative conditions including connotative impairments

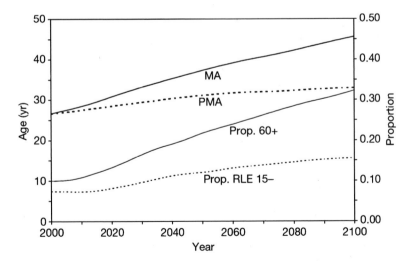

Fig. 13.3 Various projected trends in populating aging with respect to Median Age (MA), population proportion aged 60 or above (Prop. 60+), the proportion of the population with 15 or less years' life expectancy remaining (Prop. RLE 15−) and the prospective median age (PMA) defined as an adjusted median age which takes into account the increased proportion of the population within the Prop. RLE 15− band (Reproduced from [23])

such as dementia as well as physical impairments, such as osteoporosis and arthritis as well as vital organ failure. Increased longevity also increases the risk of cancer which is explained in the theory of aging which states that cancer, in a lifetime, is a statistical inevitability. This is down to the estimated 10^{16} divisions that a cell undergoes within a lifetime which leads to a build-up in DNA replication errors [24].

Indeed, ruling out morbidity from an acute source, death can be predicted to inevitably come from a cancer, organ failure or neurodegeneration [25]. The first two of these, cancer and organ failure, have the potential for onset at a much earlier age than neurodegeneration and thus these have a tendency to monopolise the time and money of research, nanomedicine included. The consequence which has been seen as a result of this, and is being seen at increasing levels, is that an increasing volume of the population is surviving previously terminal encounters with such conditions. Neurodegeneration can be seen as one of the final obstacles in the way of humanity achieving true hyper-extension of the natural life span. As a demonstration of the scale of the issue, it is estimated that six to seven million persons in Western Europe are currently affected by Alzheimer's disease, the main cause of dementia. This number doubles every 5 years within the over 65 age bracket and has an incidence of one in three for those at and above 80 years of age [26].

Within the past decade, nanomedical research has been extremely fruitful. Ground-breaking treatments for all ailments have heralded the move towards a time where humanity will be true masters of their existence in the sense that disease in the convention that it is known today will no longer exist. Indeed, parties in favour of this future eagerly anticipate and speculate about such a time. In this sense, nanomedicine as a concept, albeit indirectly, will force mankind to re-evaluate what it is to be human. It can be argued that nanomedicine as the tool or mechanism will accelerate technological progress to such a level as could affect humanity in a manner so drastic that post-nano mankind would be unrecognisable to the minds of today.

Nanomedicine could eventually chaperone a transition from a definition of humanity that is predominantly based on the physical to more of a disembodied ideal. This will first be confronted when the level of medical care is reached to make it possible for the person, which must be defined as the consciousness within the physical body, to be allowed to exist longer than the flesh in which it is contained. The 'wearing out' of the physical body is clear to see in the elderly. In some ways, the separation of mind and body is quite obvious; for example, the development of a cancer in the abdominal cavity. In this scenario, the person is unquestionably the same person as they were before the tumour. This definition blurs significantly when this 'wearing out' is observed in neurodegenerative diseases. As will be discussed in further depth below (Sect. 1.3.6), the consciousness of a person can be generally isolated to the brain, which is the epicentre of a person's being. Once the brain's neuronal tissue begins to degenerate at the end of their life time, portions of that consciousness are lost with the degradation of the physical tissue to which it was bound.

Because regeneration of neuron cells does not occur in the majority of the adult brain, it can be argued that the consciousness of that person, which is developed over their lifetime, is inextricably bound to the flesh and thus when the flesh degrades at the end of its lifespan, so too does the consciousness. Indeed, this degeneration occurs throughout the adult life at a rate of 80,000 neurons a day to the extent that, at 80 year of age, the accumulative loss of neurons amounts to 10 % of the brain's original capacity [26].

The advances made by nanomedicine, either through repair or replacement will cause the artificial prolongation of said consciousness longer than the original flesh to which it was once connected. As the flesh is the dominating identifying feature of all living things, the application of nanomedicine to this extent will force what will be for some an extremely uncomfortable discussion as to what it is to be not only a human but a disembodied consciousness.

13.3.1 Nanomedicine: The Cure for Death?

The only things certain in this life are death and taxes. The words of Benjamin Franklin have been reiterated for several centuries and have been proven to be true whenever one cared to test them. Indeed, death was a certainty for all living things long before the initial utterance of Franklin's zealously quoted line. Due to its inevitability, death has always held a revered position within the human psyche alongside which has been placed the desire to overcome it. One of the first high profile casualties of this quest for immortality was Qin Shi Huang, the first emperor of China circa 200 BC who discovered his mortality shortly after ingesting mercury pills ironically intended to endow the immortality he so sorely wanted [27].

In this more modern time, an increased understanding of the physiological mechanisms of aging as well as the advent of transplantation surgery and bionics has made an increase in longevity much more attainable. It is without doubt that nanomedicine will have a substantial role to play in this process however, what can be called into question is to what extent will the 1,000 year old man, as heralded by Professor Aubrey De Grey to be born in the next 10–20 years [28], resemble today's perception of 'man'.

Modern medicine has adopted an increasingly mechanistic stance towards the body in the sense that individual parts can be removed, repaired or replaced. These procedures have been executed with increasing ease with respect to the passage of time, which has been facilitated by increasingly sophisticated surgical tools and autonomous surgical aides. The idea of transplantation is by no means a new one – Saints Cosmas and Damian, patron saints of surgeons, are said to have removed and replaced a cancerous leg of a Roman Deacon in the third to fourth century [29].

Organ replacement and transplantation is a procedure with a similar cornucopia of legends. An example of such transplantation can be seen in the story of the third century BC physician PienCh'iao who performed not only an organ transplant but an organ exchange. The motivation of such a procedure was to balance the constitutions of the two patients since one was strong of will but weak of spirit and the other was strong of spirit and weak of will [30]. Unfortunately for PienCh'iao, he did not publish in a peer-reviewed journal and thus, the validity of such a procedure must be speculated. The first verified human to human heart transplant was successfully completed some twenty-three centuries later in 1967 by the South African doctor Christiaan Barnard, though the patient died just 18 days later due to complications arising from the obligatory immunosuppression medication administered to guard against organ rejection [31]. Organ rejection was not an issue for the first kidney transplant which was carried out in 1954 between identical 23 year old twins Ronald and Richard Herrick who, as it turned out were ideal candidates for transplantation surgery being immunologically identical and thus absent of the need for immuno-suppression [32].

In this modern age, built upon such pioneering actions, this technique of replacing a 'faulty' organ with a healthy one from a donor is now common place with 3,740 organ transplants taking place in the UK in 2011, a record high from previous years [33]; a significant hurdle to this transplantation process is encountered when a suitable match is searched for which is most often contingent on the death of a secondary individual which is then used as the donor. Even at this point, there is still chance of the organ being rejected by the receiver's immune system, as was encountered by Barnard immediately following the first heart transplant, which requires the recipient to take immunosuppressants for the rest of their lives to protect the donor organ through the inhibition of the host immune responses [34].

The consequences of extended immunosuppression can be in intuitively deduced as a decreased resistance to infections and malignancies; a side-effect which has serious connotations on the quality of life of the organ receiver. Even so, the alternative to a transplant for most is organ failure which results in death or reliance on artificial organ assistance and substitutes most commonly recognised in the use of an implanted pacemaker, kidney dialysis, a colostomy bag, respirator use or an artificial larynx. Substitutes have been invented for each of the many components within the human body, all of which seek to restore some quality of life to the patient in the face of the failure of their natural organ. All of these, including those mentioned above, are pale comparisons to the physiological marvels that they attempt to replace.

In this nano-age, medical technology has reached a level which was previously only conceivable on the pages of science fiction publications. This is the emerging world of regenerative medicine. This facet of medicine has long been embodied in the 50 year old application of 'skin farming' to grow epidermal grafts for burn victims [35]. In more recent times, increasingly sophisticated laboratory techniques have facilitated the *ex vivo* synthesis of various more complex organs; already, basic organ and tissue analogues have been reported such as of bladders and urethras [36, 37], as well as more complex organ structures such as hearts and neural tissue [38, 39] and even, most recently, the synthesis of functioning human livers from pluripotent adult stem cells [40].

What must first be noted from these cited breakthroughs is the striking decrease in the separation time between them. This is to say, if it took over 40 years to make the transition from straightforward epidermal tissue cultures to a simple organ such as a bladder, why did the transition from a static and relatively simplistic organ to a complex and active organ such as the heart only take a further 2 years? The answer comes from the accelerative nature at which technology advances – from the shoulders of giants indeed! New nanomedical techniques such as nano-manipulation, for example, vastly increase the speed at which cellular and even intracellular handling can be done. Techniques such as cell identification and sorting are pivotal steps within the process of organ regeneration; the location and isolation of the viable stem cells of the original organ must be carried out. These must then be introduced onto a prefabricated scaffold and cultured in order to ensure their successful adherence and growth. Efficient cellular and intracellular manipulation has been made possible or enhanced through techniques and tools birthed by nanomedicine. A prototypical example of these techniques can be given by the optoelectronic tweezers which can facilitate the manipulation of cells with a rapidity and resolution never before seen [41, 42]. Armed with such advanced tools, a greatly enhanced understanding of cellular behaviour can be achieved; an understanding essential for the complexity that is implied by the growth of a complex organ structure.

The advantages of being able to 'regrow' an organ from a patient's own organ stem cells lies in the elimination of reliance on immunosuppressants to guard against rejection which in turn will increase the life-years of the patient. Indeed, these advances in organ growth have occurred parallel to breakthroughs in antimicrobial nanoparticle treatments the prototype of which can be seen in the use of nano-silver. The novel use of nanosilver as an antimicrobial is most notable in its application as a wound dressing which have been seen to be effective treatments of otherwise fatal burns and likewise could be used to protect the immunosuppressed from otherwise fatal infections [43].

To complement the progression of organ regeneration and transplantation, there has also been seen an increase in the complexity transplantations successfully executed. Hand transplantation is now relatively commonplace [44]; building on this, the first double leg transplant was also carried out in July 2011 [45] whilst the first full face transplant took place in March 2010 [46] with partial facial transplants having been carried out successfully for half a decade [47]. The increased complexity and precision demonstrated in these landmark procedures are made possible through advances in surgical tools and most apparent in the increased complexity of the robotics used in such procedures.

13.3.2 Cellular Surgery and Beyond

Enhancement of 'conventional' surgeries, taken as manipulation of tissues, can be brushed aside as the mere infancy of nanomedicine. Intracellular surgery is also becoming a reality through the application of nano-procedures. Given that the regeneration of an organ is now becoming a reality through the use of a subject's stem cells, the next hurdle to overcome will be the complete halt of cellular aging which can be defined as the shortening of stem cell telomeres, located at the end of each chromosome, which occurs during every cell division. A possible solution for the slowing of this aging process as proposed by conventional medicine can be taken as the isolation of the resveratrol, the speculated 'anti-aging' ingredient in red wine [48]. Assuming that this compound does have such an effect, it merely has a retardation effect and does not completely halt aging with respect to the passage of time. Nano-surgery, which is carried out using nano-tipped needles, can probe into the interior of living cells without causing damage to the cell itself [49]. The connotations of such a procedure will again force a radical re-evaluation of convention. While still in its infancy, this technique has been cited as to be used as a great aide in the study of cellular diseases and other cellular components [50]; to bring this idea further, there has already been synthesised artificial human telomere 'nanocircles', designed initially to study the aging process more closely [51].

It is quite plausible that this technique could be carried out ahead of the organ regeneration procedures that were discussed previously. In this manner, theoretical immortality could be achieved in the sense that repairs and replacements for the various corpus components could be manufactured and executed using a subject's stem cells with artificially elongated lifespans. This being said, whilst the idea of cellular surgery is with nano-needles is impressive, even now there is an increasing possibility that this will seem to future generations' eyes as the idea of a witty talking servant-robot seems to this generation (Sect. 13.1). This is because the idea of cellular surgery using a nano-tipped needle is an approach from the top-down which is one of the established and time-proven conventions of technology to date. As was discussed is Sect. 13.1, a shift in the paradigm of medicine is occurring at present embodied by the bottom-up approach. In the case of regenerative medicine, an example of this is the reprogramming of adult cells to their pluripotent and totipotent forms *in vivo* as was reported by Abad and colleagues in September 2013 in *Nature* [52].

Just as it has been seen within the field of particle synthesis, while the top-down approach for nanomedicine is an uncontestable revolution in itself, it is also, at its heart, an extension of existing top-down conventions. The true revolution, in many ways, is still yet to come in the form for the full implementation of bottom-up approaches to the clinical setting.

To use an example within the field of regenerative medicine, the manipulation of cells *in vitro* has already been reported and has been exploited to tailor-make replacement body parts within a clinical setting with the use of the patient's own stem cells thus eliminating the possibility of rejection that plagued conventional organ transplantation. This is embodied in the landmark success of the *in vitro* growth and subsequent transplantation of a tracheobronchial section into a patient in 2011 which was made using a synthetic tissue scaffold and multipotent adult stem cells [53]. While this is extraordinary to see happening in a clinical setting, the report of the *in vivo* manipulation of cells in animal models to reprogram them to an embryonic state, as stated above, has the potential for implications that could dwarf that entailed by the advent of tailored body parts and organs. When such techniques come to clinical fruition, it may be possible to dispense with surgery as it is known today altogether opting instead to reprogram the 'malfunctioning' cells *in vitro* and likewise to have patrolling nano-doctors (Sect. 13.3.3) to detect and prevent disease and other ailments as they occur.

Indeed, if these land mark procedures were not astounding enough, attention must also be drawn to the more bizarre transplantation procedures which have been completed in the past. Experimentation centred around the transplantation-of-consciousness hypothesis which was carried out in the mid-twentieth century in both America and Russia was seen to be largely successful [54]. In these experiments, the heads of animal subjects were transplanted onto secondary bodies, following the reattachment of the blood supplies, the re-embodied head was seen to regain consciousness and respond to various stimuli thus providing preliminary proof that such a drastic transplantation was possible. The heads of the American and Russian Laboratories, Doctors Robert J. White and Vladimir Demikhov respectively, were greeted with mixed responses to their landmark work. Undoubtedly a good number of these negative responses were due to the drastic contrast between the level of practiced medicine of the day and the complexity which head transplantation implied.

Indeed, as was previously discussed (Sect. 13.2), it is often the case whereby technological research and advancement is obliged to wait for the public perceptions to catch up. An example of this can be taken as either gene or stem cell therapy whose progress experienced notable retardation through financial starvation. This was due both to the ethical grey area which these pioneering therapies had unearthed but also, in the case of gene therapy, due to the death of clinical trial subjects [55, 56]. In the case of the head transplant procedure; the key feature which was missing from the original experiments was a mechanism of reattachment of the spinal cord thus completing the interface between the head with the donor body.

Incorporation of nanomaterials into neuronal therapy has led to the development of techniques for the regeneration of injured tissue. Growth factor-impregnated scaffolds made from aligned polymer nanofibres have been seen to induce rapid regrowth of the spinal cord tissue. The use of this technique could be highly advantageous for such a procedure as head transplantation because within its primary growth-promoting role it also suppresses the formation of scar tissue, an occurrence that severely hampers any further attempt at reattachment [57, 58]. The employment of magnetic nanoparticles has also been proposed as a mechanism for reattachment of spinal cord tissue. The use of the mechanical force provided by the nanoparticles would also void the former dependence on growth factors [57, 59]. It must be noted that such innovative and revolutionary proposals have occurred quite recently, as the date of the cited publications will demonstrate, and are still very much in development; however, they do lay the foundation of a surge of growth in the sector of central nervous system regeneration which will play a key role in humanity's route to immortality.

The concept of head transplantation must not be dismissed as something that could never occur. It must be noted that less than a decade ago the idea of clinical use embryonic stem cells was still somewhat of a taboo. Exhaustive promotion from the late Christopher Reeve seemed, in the eyes of some at least, to be in vain. There was however, a shift in public perception which was coaxed by a greater understanding of what the therapy entailed which was brought about by such campaigning. This education of the public and subsequent change in governmental policy has led to the first embryonic stem cell clinical trial being executed in October of 2010 [60]. The acceptance of this preliminary trial achieved somewhat of a flood gate effect. There are now several trials using embryonic stem cells throughout the world many of which are achieving notable success; Success which can be personified in the UK based stem cell trial for the treatment of Stroke victims [61]. This trial achieved its initial goal with enough authority to merit its extension which was announced at the start of September 2011 [62]. As of September 2013 the company reiterated its commitment to its targets and projected the start of phase II trials for the aforementioned treatment, codenamed ReN001, as well as the commencement of phase I trials of second treatment, codenamed ReN009, for the vascular condition ischemia by the end of the year [63]. The acceptance of stem cell therapy could be a very good model with which to project the acceptance of the radical procedures that nanomedicine will lead to, in the same way that the success of the ReNeuroncan be an indication as to the potential rewards for the future of healthcare which await such an acceptance.

To return again to the idea of head transplantation, in terms of the complexity that such a procedure would entail one must only look to the progress made in the human genome project. The project, having started in 1990, completed an initial 'draft' in 2000. A further 3 years saw the announcement of the true completion of the human genome [64]. What followed could be likened to the opening of a flood gate for research whereby it was attempted to map the genomes of each demographic. This flood of research has progressed to such a stage that, as of the end of 2011, there was an estimated 30,000 completed human genomes [65].

It must be acknowledged that a procedure such as head transplantation, which is arguably becoming increasingly feasible with the medical advances discussed above, does not seem as outlandish an idea to some as one might think. The practice of cryogenically freezing the body directly following death is becoming somewhat of an accepted phenomenon. This follows the 'cryonic suspension' of the first person almost half a century ago; one James Bedford, a professor of psychology and well-known advocate of the then-theoretical technique. As could be expected, the speculative preservation of his body in January of 1967 was greeted with controversy and several court cases [66]. Since then, there has been an explosion of companies all of which offer this same service, for a fee which is argued to be nominal relative to the benefit it is speculated to provide. A list of which can be seen in Table 13.1.

What is most interesting in the context of this discussion with regard to this speculative freezing process is the option of 'neurosuspension' which is the preservation of the head of the subject only. In contrast to the freezing of the entire body, the

Table 13.1 List of organisations which offer cryonic suspension

Company name	Location	Foundation
Alcor Life Extension Foundation	Arizona, US	1972
American Cryonic Society	California, US	1969
Cryonic Institute	Michigan, US	1976
EUCrio	Braga, Portugal	2010
KrioRus	Alabychevo, Russia	2005
Suspended Animation Inc.	Florida, US	2002
TransTime Inc.	California, US	1972

neurosuspension option offers to suspend the disembodied head or 'consciousness' of the patient in the speculation that, in a time when technology can facilitate 'reanimation' of a patient, it would also be viable to synthesise an entirely new replacement body for the bodiless 'person' [67].

Technological advances have already progressed this reanimation process a great deal; while the complexity of reviving a brain is still beyond the reach of current technology, increasingly complex organs tissues and, more recently, whole organs are being successfully frozen and then replanted [68, 69]. The primary source of damage in this case is the formation of ice crystals within the tissue; again, there have been proposals to bypass the need even to keep the physical remains of the subject by 'downloading' the respective consciousness to a secondary location via the concept of total brain emulation (Sect. 13.3.4). However, this then forces the need to seriously debate and define for the fathomless depths of what it is to be human, an individual and perhaps even, a consciousness itself (Sect. 13.3.6).

13.3.3 Nano-Doctors: The Perpetual House Call

The concept of *in vivo* 'nanofactories' (Fig. 13.4) has also been put forward as a future direction of medical research. These factories, described as pseudo-cells, would exist *in vivo* in order to detect and diagnose microbial infiltration, malignancies or structural damage as it occurs. These structures would then synthesise an appropriate therapeutic in situ utilising ambient physiological molecules to do so [70]. This concept builds on many of the present novel functions of nanomedicine such as targeted delivery, stimulus-dependent execution of function and monocellular specificity. The advantage of these systems would be the prevention of illness or structural damage through the removal or neutralisation of the threat before it can manifest its conventionally recognisable symptoms.

Steps towards the realisation of this programmable nanofactory can be seen in the production of a 'synthetic life' within the laboratory of Craig Venter [71]. In this experiment, bacterial cells were reprogrammed with synthesised DNA to the end that the bacteria would then execute the predetermined function within its environment.

It must again be acknowledged that achieving such a landmark as the creation a synthetic cell, a feat that for so long was confined within the realm of science fiction, has been on the technological agenda for some time now with tireless research and development being carried out to achieve it. The progress that could be expected within the field of synthetic cell design could be expected to be analogous to the exploitation of viral protein cages as encapsulation and delivery entities, which can be taken as a breakthrough along the same vein and preceding that of the manipulation of a bacterial cell to carry out a specified function, is a technique over a decade old [72]. Following the initial breakthrough, the usage of the technology quietly grew within its primary niche, gene delivery, to such an extent that as of 2011, it was being used in approximately 70 % of the 1,714 gene therapy clinical trials underway worldwide [73]. Indeed, such progression from manipulation of a simple viral structure to

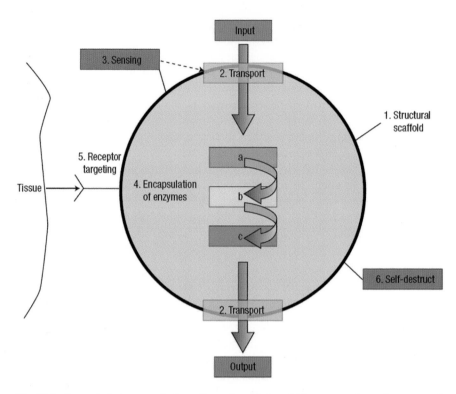

Fig. 13.4 Conceptual structure of a 'nanofactory' consisting of (*1*) a structural membrane or scaffold, (*2*) a mechanism for transportation of biomolecules in and out of the factory interior, (*3*) a sensory application, (*4*) biochemical machinery or enzymes for compound manufacture, (*5*) a targetable ligand attachment for targeting within the body, (*6*) an inherent self-destruct mechanism triggered upon completion of the nanofactory's primary goal (Reproduced from [70])

prokaryotic cell can be used to extrapolate the time to progress from prokaryote to complex eukaryote. The 12 years between these two landmarks could be expected to short relative to predictions, given the experience of the previously discussed human genome project (Sect. 13.3.2) in terms of rapid acceleration in progress following the achievement of the initial goal.

13.3.4   Nanomedicine and Prostheses

The ultimate melding of technology to medicine can be found in the form of increasingly sophisticated interfaces between man and machine. This has long been the goal of prostheses. Two of the main limiting factors within this quest can be given as intricacy and deftness. While each one is nigh on synonymous with the other, the pair can be separated thus: intricacy implying a complex interface, such as a device 'plugging in' to the brain and deftness signifying the extent to which the Darwinian perfection of a biological organ or appendage can be mimicked.

An example of this deftness can be found in one interesting and most recent development which was announced in May of 2013. This was the approval of clinical trials of a biosynthetic hybrid artificial heart, the most advanced of its kind to date [74]. This approval comes approximately 4 years after the granting of the patent rights to this design [75]. The incorporation of a biological element within this synthetic device comes in light of the increased manner in which biological tissues can be manipulated due to technological advances and an increased understanding of the behaviour of such materials as discussed in Sects. 13.3.1 and 13.3.2. In this instance, the use of such 'living' materials reduces complications experienced by previous forays into bio-mimicry such as blood clot formation and device rejection by the host. The manipulation and incorporation of, in this case, animal tissues in to a clinical device can be taken as an indication as to the direction in which this particular field of prosthetics is headed.

In relation to intricacy, there is no organ with a greater inherent intricacy and importance than the brain. It has become possible to interface directly with the brain or nervous system so as to vastly increase the utility of exterior prostheses: eyes and ears for example. The argument in favour of such prostheses has always been foiled due to their lack of sophistication in contrast to the biological component they attempted to mimic, caused predominantly because of the barrier which stood between the user and the appendage. This was due to the stark contrast between the then-current state of the art and computing power as well as electricity required to process the massive amount of data which a biological appendage amasses through its intricate function.

A direct interface with the brain need not be as futuristic a concept as may be initially thought. Indeed, both cochlear and sight-enhancing implants are both well-established concepts. Cochlear implants are perhaps the most basic methods of interface with the brain and certainly one of the least invasive or, at least, one of the most accepted [76]. Their wide spread use can also be contributed to their comparatively basic technological requirements relative to the other senses.

Visual aids implanted directly into the visual cortex can be viewed as something of a threshold because, in contrast to the cochlear implant, which is interfaced through the cochlear nerves within the ear, or indeed the research into interface with the optic nerve [77], such a visual aide requires interfacing directly with the brain. This idea, again, sits upon comparatively ancient roots with the first attempted cortex-camera interface being attempted in the early 1970s [78]. This initial attempt, as one could imagine, was a long way off the capabilities of the biological eye, nonetheless, it was a decisive start. Though there has been notable progress in the time between then and now, the attempt of interfacing such a complex network as the visual cortex with an artificial component highlights the limitations of non-nanotechnology based systems. This limitation is the phenomenal intricacy with which the many tens of thousands of neurons are connected within the cortex [79]. In order to emulate a system with such complexity, a radical rethink must occur at the fundamental level of established computing convention.

This radical rethink is already in motion today with nanotechnology as its primary driver. In order to recreate the exquisite detail and efficiency of the human or indeed any biological entity with cognitive function, an overhaul in the base hardware units of computer processing must occur. An artificial synapse connection has been synthesised with the use of carbon nanotubes [80]. This can be hailed as the beginning the design and synthesis of more complex brain prostheses which in time could be part of routine surgical treatment for degenerative brain conditions. In parallel to this development is the work to replicate hippocampus structure in which would serve to save the memory of a person which would have been lost due to neural degeneration [81]. This breakthrough, perhaps more so than any other has potential connotations in lieu of its proposed application to essentially 'store' a portion of an identity within it. Subsequent to its announcement in 2003 where it was hailed as 'the world's first brain prosthesis', further work carried out proved the concept such that the artificial chip could cause recollection of a predetermined function in the animal models used [82].

Complementarily to these initial animal models and hailing another major jump into the future of medical treatments, actual clinical brain interfacing has also been established for some years. The first direct interfacing device which read an output from the brain was achieved in 2004 when a severely paralyzed patient was interfaced with a small hundred-electrode sensor which interpreted the signal in a manner which could then be used to direct a cursor around a computer screen [83].

Mediation of brain signals has also become an accepted clinical practice for the treatment of 'faulty' brain function arising from debilitating neuronal disorders personified by conditions such as Parkinson's disease or through traumatic brain injury. Though deep brain stimulation for the inhibition of symptoms, as is the case for Parkinson's, has been well established having been introduced more than 20 years ago [84], behavioural modification through the implantation of a brain 'pacemaker' has also been shown quite recently [85]. With the implantation of such a device, many quality-of-life functions can be returned to the patient which were previously been lost through injury. This artificial interface is also being proposed from non-neurodegenerative conditions such as depression and schizophrenia [86, 87].

What is being observed in the progression of the research cited above is a movement towards a sophistication of medical treatment which undoubtedly will not be able to progress without the aid of nanomedicine. To take the goal of total human brain emulation as an ultimate goal, the essentiality of nanomedicine as a chaperone becomes clear. The human brain can be viewed as the ultimate engine from which all human knowledge stems. As a physical entity, its complexity is gargantuan; consisting of a network of 100 billion neurons each of which can have a connectivity of up to 10,000 synapses [80]. The computing power of the human brain has been estimated to be equivalent to 100 teraflops (floating point operations per second) [88]. Though such high performance rates have been achieved as far back as 2003 [89], the energy requirements to achieve this, cited to be enough to power 7,500 domestic homes, are grossly inflated from that of the biological brain, generally given as approximately 20 W depending on cognitive load, which it was attempting to match. If brain prostheses are going to be a viable answer to the problems manifested by age-related neurodegeneration, incorporation of nanotechnology in unavoidable; with the trend towards neurodegeneration seen in the population now, it can be reasonably speculated that the demographic effected will grow a great deal in the coming years and thus so too will the global dependency on increasingly more complex and innovative neural prostheses.

Increased feasibility of the idea of total brain emulation has led to the publication of speculative predictions of the requirements of such a feat. The most notable of these could be taken as that published by the Future of Humanity Institute based out of Oxford University [90]. In this report, a systematic mapping of all the aspects and connotations of brain emulation is done. What is acknowledged first and foremost is the speculative nature of such report in the sense that such a reality as total human brain emulation is so far off at this point and subject to the advents and changes in technology that a true accurate prediction is not a real possibility.

13.3.5 The Tithonus Error

Just as the direct positive effects of nanomedicine can be traced back to a deep seated desire for extended life echoed on mythology, the indirect implications of nanomedicine can too be traced back to such fabled roots also. Already, the less-than-desirable consequences of having an artificially extended life span can be clearly identified. This is primarily the notable decrease in quality of life which accompanies old age. This concept is known as the Tithonus error [91]. Tithonus was a Trojan prince made famous because of the deal struck between his titan lover, Eos, and Zeus. She implored Zeus to grant eternal life to him though by forgetting to request eternal youth to compliment it, Tithonus was forced to endure eternal life without eternal youth.

As was discussed previously (Sect. 13.3), medicine has brought this generation to a point, as demonstrated by the aging of the population, at which the Tithonus error is more apparent than ever. This can be generically embodied by dementia, which can be likened to the degeneration of the very character of the person through memory loss et cetera. More so than before, an emphasis must be placed on improving quality-of-life in old age at the very least equal to that on curing diseases relating lifestyle such as cancer and organ failure which previously served to shorten life span.

One must only look at the advent and popularisation of the euthanasia movement and its subsequent legalisation in such places as the Netherlands, Belgium, the US state of Oregon, the northern territories of Australia and Switzerland [92] to see the growing need for further research into quality-of-life improvements for sufferers of chronic and debilitating conditions, as discussed below, as well as a greater acknowledgement for the need for further debate for such treatment or action as euthanasia. Indeed, the reality is that with such boarder-dissolving agreements as were made to create the Schengen zone in mainland Europe, it is becoming more and more complicated to enforce an individual nation's laws regarding such an issue as euthanasia. The ability to travel from a country where such a procedure is still illegal, such as the UK (which incidentally is outside of the Schengen zone though a member-state of the European Union), to a country permitting it, such as Switzerland, was documented by the celebrated British author and diagnosed Alzheimer's disease sufferer Terry Pratchett in the controversial 2011 BBC documentary 'Terry Pratchett: Choosing to Die' [93]. The creation and airing of such a program demonstrates that those with a will to do such a thing could, can and do indeed do it, with or without the approval of their native country as well as the growing awareness of such a practice as euthanasia in the public psyche.

The end of the first decade of the millennium marked the start of the arrival of the Post-World War II 'Baby Boomer' generation to retiring age. With this landmark passed, the world population is set to undergo a dramatic change in demographic. Between 2000 and 2009, the proportion of the world population over 60 increased by 14 %. Indeed, this has followed a tripling of this demographic from 1950 to 2000. Extrapolation of this trend projects a further tripling of this demographic between 2000 and 2050 [94]. This is expected to send dependency ratios, which is measured by the volume of the population aged above 65 and below 15 years versus the proportion of the population within the 'working' band of 15–65, as high as 70 by 2050 [95].

Chronic illnesses, which can be taken as inevitable in an increasingly elongated lifespan, can be divided into three categories. These can be given, as mentioned previously in this section, as cancer, organ failure and dementia [25]. The first two, cancer and organ failure, have received a majority of the attention from nanomedicine research due to their malicious track record of dramatically shortening life spans. While innovation stemming from nanomedicine has caused a greatly increased chance of overcoming these conditions, the repercussions of overcoming these previously terminal illnesses will be profoundly far-reaching. Such a triumph of modern medicine over these illnesses will cause, and to an extent has already caused, an upheaval in medicine towards increased emphasis on age related ailments which have chronic rather than acute onsets.

Such an upheaval in medicine has been a long time coming; palliative care has long been the sole consolation to those afflicted with the Tithonus error; though as stated above, euthanasia, though somewhat of a taboo, has been practiced in throughout history as it is being practiced in this day and age. Palliative care is not something isolated to the comparatively recent advent of nanomedicine; the hospice movement which can be linked to the tripling of the over-1960s demographic between 1950 and 2000 having been started at approximately the same time [96]. In pre-twentieth century medicine, the need for such palliative care was limited due to the shortened life expectancy of the time. The twentieth century saw the advent of better and more sophisticated practices through greater understanding of the diseases and conditions being treated. In hand with this, came medical advancement and new techniques and treatments such as radiotherapy [97], chemotherapy [98] and the first mass-produced antibiotic, Penicillin [99] all of which came to be in the first half of the twentieth centuries.

The acceleration of medical technology within the twentieth century increased numbers requiring such palliative care to the level that provoked the commencement of the hospice movement which has been carried on since then to the present day. At this point, the advent of nanomedicine has begun a revolution in healthcare not dissimilar to that seen in the twentieth century in the sense that it will remove the death sentence that were for so long associated with certain diseases. This will, again, cause an exponential growth within the medical field in the manner which originally provoked the aforementioned movement; the difference being that in this modern age, mere palliative care will not be an acceptable option. Instead, medicine must embrace the radical change in perspective which will undoubtedly be greater than anything that has been seen before as well as dealing with the ethical dilemmas that will be pushed to the forefront because of it.

13.3.6 A Brave New World of Nanomedicine

Undoubtedly there will be logistical and economic problems that are associated with an increasingly long-lived population stemming from increasing demand and dependency on finite resources as were foreseen in the Malthusian population model, as too will there be ethical issues surrounding the increasingly radical treatments that will be required to ensure the aging populous has a quality-of-life justifying their prolonged longevity and indeed, the radical interventions that may be deemed necessary if this quality-of-life cannot be achieved or sustained namely, the option to choose death rather than a prolonged life of insufficient quality (Sect. 13.3.5) There are, however, far greater connotations that nanomedicine will be involved in as a result of its increased acceptance and application. The most drastic of these stems from the increasingly successful treatment of neural conditions and neurodegenerative diseases via artificial means.

The concept of brain 'pace markers' has already been discussed above (Sect. 13.3.4); while still in their infancy, they are predominantly used as mediators of basic function. This is implicated in their use to treat Parkinson's disease by way of dulling the bodily tremors that are symptomatic of the condition. This form of neurodegeneration does not impinge on the personality of the person and thus can be isolated in the realm of physical debilitation directly caused by a neurological degeneration. Proposals for the treatment of depression and schizophrenia, conversely, can be taken as disorders far more related to genetics and personality of the individual [100, 101]. Given that these proposals are occurring at the relatively young age of the technology can be taken as a sign of the potential scope of such a technique. Personality and behaviour modification or modulation through interface with an electronic controller has the potential for a striking improvement of quality of life for those suffering debilitating conditions which may or may not be related to age. The intimate connection that such an electronic device provides between the brain and the controller, particularly with the intricacy being allowed through nanomedicine, has the potential for invasive mind control, be it intentional or not. Of course it is true that the same can be said on some level about any existing psychoactive drug but the introduction of an electronic device into the brain is a step beyond the ingestion of a pill since there is the potential for far more direct control of the targeted aspect of the subject.

Progressively more complex proposals of 'mind control' are being proposed as solutions to behavioural disorders. Indeed, promising results have been yielded from the use of deep brain stimulation to combat eating disorders and alcoholism [102, 103]. The ability to artificially control such powerful behavioural traits has implications which have been heralded for quite some time. Indeed, it was a central theme in Anthony Burgess's 1962 dystopian novel 'A Clockwork Orange' in which the protagonist was conditioned to feel violently ill when confronted with violence – or classical music – so as to render him unable to be in the presence of violence via virtual incapacitation; when this secret was found out by one with motivation to harm the protagonist, this conditioning was then used in a malicious way to his apparent-detriment [104]. Willingly surrendering control to a secondary entity, be it behavioural conditioning or an implanted computer chip puts one at the mercy of those who hold the controls. This sentiment is superbly represented in Author Koestler's words [105]:

Who is to control the controls, manipulate the manipulators?

Indeed, this is not limited to the potential of direct behavioural manipulation; such a question arises with the proposed use of 'nanofactories' whereby a level of intelligence is bestowed upon an entity which is then entrusted with a colossal responsibility: the health of the subject. More so than ever before, there will be potential for mankind to do increasingly deistic acts the repercussions of which may not be seen for some time afterwards. A model for such repercussions can be seen in the emergence of nanotoxicity (Sect. 13.5). A very fine balance must be struck between caution and optimism. Given the cautionary tales of the past as well as the potential for harm which nanomedicine possesses, each step taken in the progression of nanomedicine must not be taken lightly.

Less novel questions must also be confronted with new vigour in the face of the accelerative effect which nanomedicine is having on medical research. This is namely, what does being an individual or indeed a human entail? The idea of a soul may seem to some to be out of place within a scientific document but certainly, the idea of the personality of an individual is a concept which is universal.

In this age of modern medicine, there is an increasingly mechanistic view of the human body in the sense that, if a component is not functioning in a satisfactory manner, it can be repaired, removed or replaced. This attitude of medicine can trace its origins back to and beyond the pioneering work of the renascence anatomists which encouraged its students the view of the body less as the 'image of God' whose defilement could be thought to be sacrilege, and more as a complex mechanism of interdependent components [106]. It is quite interesting to contrast the interpretation of ancient cultures of the body-whole as the 'seat of the soul' to modern attitudes of medicine in the sense that it has not been until quite recently that the mind within the head has been qualified separately from the rest of the body. The roots of this distinction can be recognised in the words of Juvenal, a roman poet who wrote in the early second century:

Menssana in corporesano

A healthy mind in a sound body. Even with this initial separation, there has always been an inherent link in terms of a person's identity with the physical body. With transplantation surgery becoming increasingly ambitious, as the concept of a full face transplant (Sect. 13.3.1) would indicate, medicine seems on some level to be distancing itself from such an idea. Even so, it cannot be denied that it is common for transplant recipients to experience some manner of change in personality traits in keeping with the foreign organ that has replaced their own [107–109] thus renewing the argument in favour of the ambiguity of the location of a person's 'soul' or consciousness.

The degree to which this testimony will impact the momentum with which transplantation and regenerative medicine is moving is undoubtedly small; however with the increase in population age will come a corresponding increase in the incidence of transplantations. It is a reasonable estimate that by 2050, the science of organ regeneration will have been greatly advanced if not perfected through nanomedical progress. At that point, the question as to how much a personality could truly be affected by an increasing number of replaced or regenerated organs or body-parts perhaps will perhaps be easier to answer, until then only speculation is possible.

More interesting even than this will be the behaviour of the first recipient of an artificial memory implant as has been cited previously (Sect. 13.3.4) whereas the exact location of the seat of the soul if not its entire existence is debatable, the hippocampus as the location of human memory is not. When the first memory implant is implanted, it will mark a profound step in human evolution toward disassociation from the physical body the likes of which has never before been seen. The personality of a person is so much based on the flesh and blood that the implications of the separation of the mind from the body challenge the very notion of human consciousness itself. Indeed, with the report of the successful deletion of targeted memories [110] and the announcement of the creation of false memories through direct manipulation of neurons in animal models, it is not unreasonable to foresee a time where memory, personality or behaviour of a subject could be altered at will until the desired effect was attained. This, coupled with the connotations of artificial implants such as were earlier discussed, is nothing short of a Valhalla for conspiracy theorists.

The next step, if not the definitive answer to this, will come with the first brain emulation whereby a new definition of life must be created to avoid complete indistinction from a non-living computer programme. This will be the ultimate realisation of the philosophical concept first described by Descartes as 'Dualism', or the relationship between mind and matter. Whether a mind can exist in the same manner outside of its physical body has been one of the most perplexing questions in philosophy since its proposition in the seventeenth century. This concept has been challenged by various futurists throughout the years with various degrees of scepticism; perhaps the most relevant to this discussion is the vision of Masamune Shirow portrayed exquisitely in the publication of 'Ghost in the Shell' [111] in which human-computer interfacing has blurred the definition of humanity to the degree where a self-aware computer virus could request diplomatic immunity. Of course, such a scenario is a great way off but it is the privilege of this generation to be present at the dawning of the age of such significant scientific discoveries.

13.4 The Boy Who Cried … Self-Assembling Semiconductors!

As is the case in so many emerging technologies, it is extremely easy to allow one-self to get swept away by fantastical speculation. This is made all the more relevant given the topic of the preceding section and the seeds of hyper-expectation it sows. It must be acknowledged in the face of such grand predictions that nanomedicine is merely a facilitator or a means to achieve a step forward in science. In the words of Thomas Edison:

> Genius is one per cent inspiration, ninety-nine per cent perspiration

It is the nature of humanity, be it industrial, commercial or the public, to specu-late on any up-and-coming market or technology with the potential to yield a profit-able end. This mentality has littered history with the remnants of burst bubbles; relics of man's fondness for snake oil, which were superimposed over emerging markets throughout history. Within this generation there have also been several; these are most commonly recognised in the dot com bubble of the late 1990s or the more recent world banking crisis of 2008 respectively, the implications of which are still being felt to this day. The root-cause of all of these market plunges is an inher-ent hyper-expectation which causes an over confidence in the speculated progres-sion and growth of a trend, be it technology or purely financial. Scandal as a consequence of hyper-expectation has also occurred within nanotechnology and, though it was spared the magnitude of financial grief seen by previous and subse-quent bubbles, the academic turmoil it caused has given harsh perspective to any overly zealous speculators in the field.

This scandal occurred at the turn of the millennium. In the late 1990s, Jan Hendrik Schön, a newly graduated and highly promising physicist doctorate, was awarded a position in the world renowned Bell labs. Bell labs' reputation can be indicated by the 11 Nobel laureates associated with the establishment through its history [112]. Schön's research interest centred on condensed matter physics and nanotechnology; true to the reputation of his employer, within a short period of time, Schön reported a revolutionary break through: the successful synthesis of a self-assembling semiconductor using organic die molecules. Retrospectively, the achievements which Schön were claiming had been made a full 20 years ahead of subsequent projections in the case of self-assembling nanostructures (Fig. 13.2) and close to a decade ahead of the final realisation of a molecular transistor [15]. The timing of this discovery did not trouble anyone at the time it was reported due, per-haps, to the reputation and academic weight of the establishment making the claim.

What followed could be described as nothing short of a frenzied fit of publication which lasted nigh on 3 years. In this time, Schön was published several times over in some of the highest ranking and respected journals in the world; Nature, Science, Physical Review, and Applied Physics Letters all willingly and repeatedly published Schön's work to the extent that by 2001, he was being cited as an author in a journal every 8 days [113]. Capitalising on this momentum, he continued on to claim to have made significant steps towards achieving molecular scaled transistors [114]. Such was the influence of the establishments backing his work, much of the scientific community were accepting, albeit with bemusement, that such prolific publication portfolio was the new benchmark for research and discovery within nanotechnology.

Following Schön's lead, a flurry of activity ensued as different institutions attempted to replicate the results cited in the literature. Following what was undoubtedly a vast investment of time and money from many of researchers in the field, the scandal of the century broke. Just 2 years into the twenty-first century and with nanotechnology still very much in its infancy, it transpired that Schön's work was fraudulent with much of the published graphical data being directly copied and pasted from one article to the next and originating from mathematical functions rather than experimentation. Following a formal inquiry, Bell Labs concluded that the lofty claims which had catapulted Schön to stardom were ungrounded [115]. Before this happened however, citations of Schön's bogus publications were countable in quadruple digits.

In the ensuing controversy in which Schön's explanation for his actions regarding the doctoring of experimental data was to magnify existing trends, when asked to procure this raw data, it transpired that Schön had erased it from his electronic records stating that his computer storage space was insufficient. In response to this, the there was a unanimous retraction of journal articles including seven in Nature [116] and eight in Science [117]. In total, nearly 30 articles from reputable journals had to be retracted whilst several articles had their content thrown into question and the scientific community stood on in disbelief [118].

The success of such a scoundrel to deceive the scientific community will no doubt ring hollow in the ears of researchers and potential investors alike when confronted with any further report of alleged breakthroughs within nanotechnology, in medicine or otherwise. One could quite reasonably argue that such perspective was achieved at a nominal price in comparison to the collapse of such bubble-growth in other fields. It was perhaps the rapid nature with which the situation grew and the scandal was exposed which may have saved many from more painful losses or embarrassment. The worth of such perspective early on in the growth of a field is invaluable for level headed and objective growth and investment based on accurate projections of the outlook of the sector.

With this in mind, in an analysis published in Nature in 2011 [119], it was found that the number of retractions of published articles had increased 10-fold in the preceding decade based on data extracted from Web of Science and PubMed. It is interesting to see how far reaching the Schön scandal has been for the scientific community in the sense that, though post-2005 saw the majority of the retractions coming from lower impact factor journals, the statistics for the years 2000–2005 place Nature and Science on the podium for the most retractions with approximately half the quantity of retractions issued by each coming from the Schön scandal. It is also interesting to note that of the total retractions cited, a full 44 % of these were removed due to misconduct with another 11 % removed because of irreproducible results, leaving 45 % the categories of honest error and 'other'. This would imply that while the Schön scandal was shocking and unprecedented from the point of the sheer volume of fraudulent data published by an individual in an emerging field, it could be unfortunately argued that regardless of the field, the dynamic and speculative nature of science lends itself to mistakes and null hypotheses. Thus all published results and claims must be greeted with a measure of scientific scepticism in keeping with the magnitude of the claims being made, regardless of the reputation of the institute, individual or journal making and publishing the claim.

13.5 Nanotoxicology and Regulation

A primary attenuator of realisation of the initial hyper-expectation of nanomedicine is the emerging field of nanotoxicology. This field incorporates the less desirable effects of nanomaterials within the physiological environment. Just as nanomedicine has brought a new wave of innovation and growth, the corresponding emergence of an entirely new and previously unknown form of toxicity has also been seen. Nanotoxicology documents material toxicity at a level of complexity far in excess of the simplistic and well established Paracelsus model for modern toxicity whereby the primary variable is dose alone [120].

Indeed, the nanotoxicity paradigm contains several variables within a single material alone. These include size, shape, structure, solubility and surface charge, and chemical composition [121]. The discovery and development of this subdivision of the field came in parallel with on-going research and application of nanotechnology in medical treatments. It is almost two decades since the Food and Drugs Administration (FDA) approved first generation nanomedicines (Fig. 13.2) in the form of liposomal preparations, this approval is widely accepted to have been made before a true understanding of the issues which the application of nanomaterials would present [42].

The discovery of the potential adverse effects of nanomaterials sparked a global scramble to put in place adequate regulation for the sector. This move, whilst very much required for the safe integration of nanotechnologies into medicine can be identified as a significant bottleneck for the transition of nanotechnology from concept to clinic.

With the ever-increasing scope of nanomedicine as well as the two decade time period since the clinical approval of the first nanomedicines, it might be surprising to know that the FDA until did not have guidelines specifically pertaining to nanomaterial-containing products until April 2012. It was at this point that two draft guidelines were issued for nanomaterial-containing food substances [122] and cosmetics [123]. Each of these documents, described as non-binding recommendations, were issued for public comment in the hopes that the previously broad regulatory approach, where new nanoproducts were reviewed alongside non-nanoproducts, would become a more nuanced affair with the aide of increased understanding and information as well as this requested public input [124, 125].

Similar to that in America, in the European Union the initial strategy was to neglect the publication of a clear official definition of 'Nanotechnology' instead opting for a 'broadly inclusive approach' to regulate nanotechnology under the umbrella of those existing frameworks; This situation arose due to the decision to initially define nanomaterials under the pre-existing term of a 'substance' [126] up until 2010 when an official definition was released by the European Commission (EC) [127] in which it referred to nanopharmaceutics, for instance, as being covered by the pre-clinical safety precautions of pharmaceuticals. Though the definition of what constitutes a nanomaterial is useful, the move was intended to enable more efficient regulation and application so as to reduce the waiting time for therapies to receive a verdict. In this way, any ambiguity resulting from boundary conditions for nanomaterials such as size limitations were removed [128]. As of September 2013, as was done in the United Sates, a request for public contribution and recommendation for modifications to the REACH annexes on nanomaterials in order to improve regulatory efficiency [129]. Just before this, a request for tenders was issued for a study 'to assess the impact of possible legislation to increase transparency on nanomaterials on the market' [130].

The delay to implement regulatory measures is understandable given the connotations that inadequate legislation could have on research momentum. While the decision to include nanomaterials in an umbrella definition was a good short term solution, such a decision then obliged individual nanomaterials to be reviewed on a case-by-case basis. This lack of definition also permitted certain nanomaterials to be freely sold as medicinal agents under the definition of 'food substances' thus obliging the regulatory powers to redefine such definitions with the advent of nanotoxicology.

Taking nanosilver as a case study of this, the difficulties arising from retrofit, case-by-case legislation can be clearly seen. Colloidal suspensions of silver have been freely available for over 100 years to any member of the public who cared to buy it [131]. It was originally used in a diverse range of applications such as pigments, photographics, wound treatment, conductive or antistatic composites and catalysis. It was after the advent of penicillin and modern antibiotics, that nanosilver's antimicrobial properties were used specifically as a substitute for conventional antibiotics due to its apparent lack of adverse side effects commonly associated with antibiotics. The need for stricter regulation of nanosilver only became clear when it was discovered to potentially be a significant health risk due to its high cytotoxicity [132]. The sale of these nanosilver products, however, continued regardless due to the aforementioned lack of a differentiation between silver's bulk and nanoscopic modulus and corresponding absence of its specific regulation within the public domain. This lack of definition allowed companies to sell such products as health-promoting food supplements [133]. This unchecked domestic sale of nanosilver through this food-supplement loophole was finally stopped at the beginning of 2010 with the implementation of Commission Regulation No. 1170/2009 [134].

Attempts to regulate the nanomedicine sector at such a relatively late time in its growth has resulted in a unique limbo-esque situation which has been artificially created by various regulatory bodies whereby implementation of nanomedicine into daily public life has been, to a large degree, halted or at least slowed significantly from the pace at which it was being distributed. Such attenuation has come as a direct result of the better understanding and realisation of the true power and potential of medicinal nanomaterials which has, for better or worse, dulled the whimsical ideologies which had previously accompanied conceptual nanomedicine.

13.6   Conclusions and Future Outlook

Nanomedicine has yielded technologies and techniques that have and will revolutionise the field of medicine. This has been done utilising the materials, properties and phenomena which are unique to the nano-scale. Nanotechnology has provided the means to continue the progression of computing power and sophistication in keeping with and perhaps eventually even surpassing Moore's law. With the security that such an assurance as this provides, predictions can be made as to the future state-of-the-art. This law can also be used as an example of the hyper-expectation that has become part of this modern society. This generation has grown accustomed to the rapid evolution of technology to the extent that the rapid progression that has been seen and forecasted within the field of nanomedicine can be taken as normality.

Such hyper-expectation transfers to healthcare as well. Increasingly high levels of sophistication in healthcare and treatments have brought about an increased global life expectancy which is projected to increase further with time; in this sense, nanomedicine has a direct hand in the creation of the next crisis in medicine. For many years, chronic illnesses embodied by organ failure and cancer have kept net life expectancy below a certain point. Advances in treatments such as regenerative medicine, more efficient therapy and earlier detection, all of which have intimate connections to nanomedicine, have served to greatly increase the numbers now surviving to the extremities of old age. In this way, nanomedicine is having a direct hand the aging population of the first world which brings with it a radical change in the practice of medicine towards an increased focus on geriatric conditions. This transition also brings with it new challenges to healthcare as well to society and, with the population of over-60s set to triple between 2000 and 2050, will potentially create an 'identity' crisis within the population as an increased reliance on prostheses and pharmaceutics, which will go in hand with the aging population, takes hold.

With such lofty expectations and projections for nanomedicine, it is easy to find one's head in the clouds whilst speculating the future state-of-the-art. Such a societal ethos brings with it an increased vulnerability to the proverbial snake oil salesman in the sense that, as a consequence of the pace of progress, one must often rely on what is reported in the literature, rather than what is actually directly observed, to interpret the level of technological innovation of the day. This ethos has already proven to be a risky one for nanotechnology. With the conclusion of its first major global nano-related scandal, professionals and laymen alike have been forced to view the grand plans and projections offered by nanomedicine in the harsh light of reality in order to better estimate the legitimate future progression of the field and to differentiate such estimates from the farcical claims made out of context and well before their time.

There must also be a heightened awareness of the far reaching implications of the application of nanomedicine. These implications can already be seen with the advent of nanotoxicology and the scramble to retrofit regulation to an industry experiencing prodigious growth worldwide. A delicate balance must be struck between nanomedicine as a conceptual idea and it as a practiced discipline. The regulatory bodies, which are at the helm of the assimilation of nanomedicine into wide spread use, have been charged with the intricate task of determining the safety of the nanomaterials used within these new techniques whilst simultaneously not entirely halting or bottlenecking nanomedical progress.

This generation has been granted the great privilege of witnessing the transition of society from the pre to post nano-age and from this vantage point, can be seen the melding of both conceptual hyper-expectation and the dawning realisation that comes with attempts towards implementation.

References

1. Booker KM (2006) Forbidden planet. In: Alternative America: science fiction film and American culture. Praegar Publishers, Westport, pp 43–49
2. McCurdy HE (2006) Observations on the robotic versus human issue in spaceflight. In: Dick SJ, Launius RD (eds) Critical issues in the history of spaceflight. Government Printing Office, Washington, pp 77–107
3. Brock DC (2006) Understanding Moore's law: four decades of innovation. CHF Publications, Philadelphia
4. Culver HR, Daily AM, Khademhosseini A, Peppas NA (2014) Intelligent recognitive systems in nanomedicine. Curr Opin Chem Eng 4:105–113
5. Fattal E, Tsapis N (2014) Nanomedicine technology: current achievements and new trends. Clin Transl Imaging 2(1):77–87
6. Tong S, Fine EJ, Lin Y, Cradick TJ, Bao G (2014) Nanomedicine: tiny particles and machines give huge gains. Ann Biomed Eng 42(2):243–259
7. Wiesing U, Clausen J (2014) The clinical research of nanomedicine: a new ethical challenge? Nanoethics 8(1):19–28
8. Roco MC (2005) International perspective on government nanotechnology funding in 2005. J Nanopart Res 7(6):1–9
9. NNI (2013) Coordination of the NNI. http://www.nano.gov/about-nni/what/coordination. Accessed 27 June 2013
10. Sargent JF (2013) The national nanotechnology initiative: overview, reauthorization and appropriation issues. http://www.fas.org/sgp/crs/misc/RL34401.pdf. Accessed 27 June 2013
11. National Cancer Institute (2010) Cancer nanotechnology plan. http://nano.cancer.gov/objects/pdfs/CaNanoPlan.pdf. Accessed 27 June 2013
12. Hullmann A (2006) Who is winning the global nanorace? Nat Nanotechnol 1:81–83
13. Lux Research (2004) Sizing nanotechnology's value chain. http://www.altassets.net/pdfs/sizingnanotechnologysvaluechain.pdf. Accessed 29 June 2013
14. Roco MC (2010) The long view of nanotechnology development: the national nanotechnology initiative at ten years. In: Roco MC, Mirkin C, Hersam M (eds) Nanotechnology research directions for societal needs in 2020. Springer, New York
15. Song H, Kim Y, Jang YH, Jeong H, Reed MA, Lee T (2009) Observation of molecular orbital gating. Nature 462(7276):1039–1043
16. Mansour TE (2002) Chemotherapeutic targets in parasites: contemporary strategies. Cambridge University Press, Cambridge
17. Mehra J (1994) The beat of a different drum: the life and science of Richard Feynman. Clarendon, Oxford
18. Drexler KE (1986) Engines of creation, 1st edn. Anchor Press/Doubleday, Garden City
19. Kaiser M, Maasen S, Kurath M (2010) Governing future technologies: nanotechnology and the rise of an assessment regime. Springer, Dordrecht
20. Huxley A (2007) Brave new world, 2nd edn. Vintage, London
21. Crichton M (2002) Prey. Harper Collins, New York

22. Gee JP (2003) What video games have to teach us about learning and literacy. Palgrave Macmillan, New York
23. Lutz W, Sanderson W, Scherbov S (2008) The coming acceleration of global population ageing. Nature 451:716–719
24. Alberts B, Johnson A, Lewis J, Raff M, Roberts K, Walter P (2002) Cancer. In: Molecular biology of the cell, 4th edn. Garland Science, New York, pp 1313–1362
25. Murray SA, Kendall M, Boyd K, Sheikh A (2005) Illness trajectories and palliative care. Br Med J 330:1001–1011
26. Giacca M (2010) Clinical applications of gene therapy. In: Gene therapy. Springer, New York, pp 139–282
27. Slavicek LC, Mitchell GJ, Matray JI (2005) The great wall of China. Chelsea House Publishers, New York, NY, USA
28. De Grey ADNJ, Rae M (2007) Ending aging: the rejuvenation breakthroughs that could reverse human aging in our lifetime. St. Martin's Press, New York
29. Androutsos G, Diamantis A, Vladimiros L (2008) The first left transplant for the treatment of a cancer by Saints Cosmas and Damian. J BUON 13(2):297–304
30. Wong BW, Rahmani M, Rexai N, McManus BM (2005) Progress in heart transplantation. Cardiovasc Pathol 14(4):176–180
31. Toledo-Pereyra LH (2010) Heart transplantation. J Invest Surg 23:1–5
32. Murray JE (1982) Reflections on the first successful kidney transplantation. World J Surg 6(3):372–376
33. NHSBT (2011) Statistics: transplants save lives. http://www.uktransplant.org.uk/ukt/statistics/statistics.jsp. Accessed 27 June 2013
34. Morris PJ (2004) Transplantation – a medical miracle of the 20th century. N Engl J Med 351(26):2678–2680
35. Cruickshank CND, Cooper JR, Hooper C (1960) The cultivation of adult epidermis. J Invest Dermatol 34:339–342
36. Atala A, Bauer SB, Soker S, Yoo JJ, Retik AB (2006) Tissue-engineered autologous bladders for patients needing cystoplasty. The Lancet 367(9518):1241–1246
37. Raya-Rivera A, Esquiliano DR, Yoo JJ, Lopez-Bayghen E, Soker S, Atala A (2011) Tissue-engineered autologous urethras for patients who need reconstruction: an observational study. The Lancet 377(9772):1175–1182
38. Ott HC, Matthiesen TS, Goh S, Black LD, Kren SM, Netoff TI, Taylor DA (2008) Perfusion-decellularized matrix: using nature's platform to engineer a bioartificial heart. Nat Med 14(2):213–221
39. Lee W, Pinckney J, Lee V, Lee J, Fischer K, Polio S, Park J, Yoo S (2009) Three-dimensional bioprinting of rat embryonic neural cells. Neuroreport 20(8):798–803
40. Takebe T, Sekine K, Enomura M, Koike H, Kimura M, Ogaeri T, Zhang RR, Ueno Y, Zheng YW, Koike N, Aoyama S, Adachi Y, Taniguchi H (2013) Vascularized and functional human liver from an iPSC-derived organ bud transplant. Nature. doi:10.1038/nature12271
41. Wu MC (2011) Optoelectronic tweezers. Nat Photonics 5(6):322–324
42. Jain KK (2008) The handbook of nanomedicine. Humana Press/Springer, Totowa
43. Church D, Elsayed S, Reid O, Winston B, Lindsay R (2006) Burn wound infections. Clin Microbiol Rev 19(2):403–434
44. Buncke HJ Jr (2000) Microvascular hand surgery – transplants and replants – over the past 25 years. J Hand Surg 25(3):415–428
45. Hamzelou J (2011) World's first double leg transplant performed in Spain. N Sci 2821:4
46. Barret JP, Gavaldà J, Bueno J, Nuvials X, Pont T, Masnou N, Colomina MJ, Serracanta J, Arno A, Huguet P, Collado JM, Salamero P, Moreno C, Deulofeu R, Martínez-Ibáñez V (2011) Full face transplant: the first case report. Ann Surg 254(2):252–256
47. Spurgeon B (2005) Surgeons pleased with patient's progress after face transplant. Br Med J (Clin Res Ed) 331(7529):1359
48. Baur JA, Sinclair DA (2006) Therapeutic potential of resveratrol: the in vivo evidence. Nat Rev Drug Discov 5:493–506

49. Obataya I, Nankamura C, Han SW, Nakamura N, Miyake J (2005) Nanoscale operation of a living cell using an atomic force microscope with a nanoneedle. Nano Lett 5:27–30

50. Beard JD, Burbridge DJ, Moskalenko AV, Dudko O, Yarova PL, Smirnov SV, Gordeev SN (2009) An atomic force microscope nanoscalpel for nanolithography and biological applications. Nanotechnology 20:1–10

51. Lindström UM, Chandrasekaran RA, Orbai L, Helquist SA, Miller GP, Oroudjev E, Hansma HG, Kool ET (2002) Artificial human telomeres from DNA nanocircle templates. Proc Natl Acad Sci USA 99(25):15953–15958

52. Abad M, Mosteiro L, Pantoja C, Canamero M, Rayon T, Ors I, Grana O, Megias D, Dominguez O, Martinez D, Manzanares M, Ortega S, Serrano M (2013) Reprogramming in vivo produces teratomas and iPS cells with totipotency features. Nature. doi:10.1038/nature12586

53. Jungebluth P, Alici E, Baiguera S, Le Blanc K, Blomberg P, Bozoky B, Crowley C, Einarsson O, Grinnemo KH, Gudbjartsson T, Le Guyader S, Henriksson G, Hermanson O, Juto JE, Leidner B, Lilja T, Liska J, Luedde T, Lundin V, Moll G, Nilsson B, Roderburg C, Stromblad S, Sutlu T, Teixeira AI, Watz E, Seifalian A, Macchiarini P (2011) Tracheobronchial transplantation with a stem-cell-seeded bioartificial nanocomposite: a proof-of-concept study. Lancet 378:1997–2004

54. Roach M (2003) Stiff: the curious lives of human cadavers. Viking, London

55. Marshal E (1999) Gene therapy death prompts review of adenovirus vector. Science 286:2244–2245

56. Frankel MS (2000) In search of stem policy. Science 287:1397

57. Zhu Y, Wang A, Shen W, Patel S, Zhang P, Young WL, Li S (2010) Nanofibrous patches for spinal cord regeneration. Adv Funct Mater 20:1433–1440

58. Göritz C, Dias DO, Tomilin N, Barbacid M, Shupliakov O, Frisén J (2011) A pericyte origin of spinal cord scar tissue. Science 333:238–242

59. De Silva MN, Almeida MV, Goldberg JL (2007) Developing super-paramagnetic nanoparticles for central nervous system axon regeneration. In: Technical proceedings of the 2007 NSTI nanotechnology conference and trade show, vol 2, Taylor & Francis, USA, pp 791–794

60. Mayor S (2010) First patient enters trial to test safety of stem cells in spinal injury. Br Med J 341:c5724

61. Wise J (2010) Stroke patients take part in "milestone" UK trial of stem cell therapy. Br Med J 341:c6574

62. ReNeuron Group (2011) ReNeuron receives DSMB clearance to progress to higher dose in ReN001 stem cell trial for stroke. http://www.reneuron.com/news/249. Accessed 29 June 2013

63. Nimmo J (2013) ReNeuron sticks to timeline after impressive progress. http://www.proactiveinvestors.co.uk/companies/news/61053/reneuron-sticks-to-timeline-after-impressive-progress-61053.html. Accessed 16 Sept 2013

64. Stein LD (2004) End of the beginning. Nature 431:915–916

65. Spencer N, Katsnelson A, Loman N, Hadfield J (2010) Human genome: genomes by the thousand. Nature 467(7319):1026–1027

66. Nelson RF (1968) We froze the first man. Dell publishing company (now Bantham-Dell publishing group & member of Random house), New York, NY, USA

67. Shoffstall G (2010) Freeze, wait, reanimate: cryonic suspension and science fiction. Bull Sci Technol 30(4):285–297

68. Best BP (2008) Scientific justification of cryonics practice. Rejuvenation Res 11(2):493–503

69. Fahy GM, Wowk B, Pagotan R, Chang A, Phan J, Thomson B, Phan L (2009) Physical and biological aspects of renal vitrification. Organogenesis 5(3):167–175

70. Leduc PR, Wong MS, Ferreira PM, Groff RE, Haslinger K, Koonce MP, Lee WY, Love JC, McCammon JA, Monteiro-Riviere NA, Rotello VM, Rubloff GW, Westervelt R, Yoda M (2007) Towards an in vivo biologically inspired nanofactory. Nat Nanotechnol 2(1):3–7

71. Gibson DG, Glass JI, Lartigue C, Noskov VN, Chuang RY, Algire MA, Benders GA, Montague MG, Ma L, Moodie MM, Merryman C, Vashee S, Krishnakumar R, Assasad-Garcia N, Andrews-Pfannkoch C, Denisova EA, Young L, Qi ZQ, Segall-Shapiro TH, Calvey

CH, Parmar PP, Hutchison CA, Smith HO, Venter C (2010) Creation of a bacterial cell controlled by a chemically synthesized genome. Science 329:52–56

72. Douglas T, Young M (1998) Host-guest encapsulation of materials by assembled virus protein cages. Nature 393:152–155

73. Edelstein M (2011) Gene therapy trials worldwide provided by the journal of gene medicine. http://www.wiley.com/legacy/wileychi/genmed/clinical. Accessed 29 June 2013

74. Cannet A (2013) Carmat gets approval to test artificial heart in four countries, Reuters. http://www.reuters.com/article/2013/05/14/us-carmat-artificialheart-idUSBRE94D0KL20130514. Accessed 15 July 2013

75. Grimme M, Gourgues JE, Carpentier A, Wartelle C (2009) Implantable one-piece heart prosthesis. European Patent WO2009112662 A2

76. Wilson BS, Finley CC, Lawson DT, Wolford RD, Eddington DK, Rabinowitz WM (1991) Better speech recognition with cochlear implants. Nature 352:236–238

77. Zrenner E (2002) Will retinal implants restore vision? Science 295(5557):1022–1025

78. Dobelle WH (2000) Artificial vision for the blind by connecting a television camera to the visual cortex. ASAIO J 46:3–9

79. Bock DD, Lee WCA, Kerlin AM, Andermann ML, Hood G, Wetzel AW, Yurgenson S, Soucy ER, Kim HS, Reid RC (2011) Network anatomy and in vivo physiology of visual cortial neurons. Nature 471:177–182

80. Joshi J, Zhang J, Wang C, Hsu CC, Parker AC, Zhou C, Ravishankar U (2011) A biomimetic fabricated carbon nanotube synapse for prosthetic applications, IEEE/NIH 5th Life Science Systems and Applications Workshop (LiSSA), pp 139–142. National Institutes of Health (NIH) Campus, Bethesda, Maryland, USA

81. Berger TW, Granacki JJ, Marmarelis VZ, Sheu BJ, Tanguay AR Jr (2003) Brain-implantable biomimetic electronics as neural prosthetics, Engineering in Medicine and Biology Society, 2003. Proceedings of the 25th Annual International Conference of the IEEE, vol.2, no., pp.1956–1959 Vol.2, 17–21 Sept. 2003, Cancun, Mexico

82. Berger TW, Hampson RE, Song D, Goonawardena A, Marmarelis VZ, Deadwyler SA (2011) A cortical neural prosthesis for restoring and enhancing memory. J Neural Eng 8(4):046017

83. Hochberg LR, Serruya MD, Friehs GM, Mukand JA, Saleh M, Caplan AH, Branner A, Chen D, Penn RD, Donoghue JP (2006) Neuronal ensemble control of prosthetic devices by a human with tetraplegia. Nature 442:165–171

84. McIntyre CC, Savasta M, Kerkerian-Le Goff L, Vitek JL (2004) Uncovering the mechanism(s) of action of deep brain stimulation: activation, inhibition, or both. Clin Neurophysiol 115:1239–1248

85. Schiff ND, Giacino JT, Kalmar K, Victor JD, Baker K, Gerber M, Fritz B, Eisenberg B, O'Connor J, Kobylarz EJ, Farris S, Machado A, McCagg C, Plum F, Fins JJ, Rexai AR (2007) Behavioural improvements with thalamic stimulation after severe traumatic brain injury. Nature 448(7153):600–603

86. Coenen VA, Schlaepfer TE, Maedler B, Panksepp J (2011) Cross-species affective functions of the medial forebrain bundle-implications for the treatment of affective pain and depression in humans. Neurosci Behav Res 35(9):1971–1981

87. Sesack SR, Carr DB (2002) Selective prefrontal cortex inputs to dopamine cells: implications for schizophrenia. Physiol Behav 77:513–517

88. Moravec H (1998) When will computer hardware match the human brain? J Evol Technol http://www.jetpress.org/volume1/moravec.htm. Accessed 27 June 2013

89. ASCI (2010) 100 teraFLOPS dedicated to capability computing. https://asc.llnl.gov/computing_resources/purple. Accessed 29 June 2013

90. Sandberg A, Bostrom N (2008) Whole brain emulation: a roadmap. Future of Humanity Institute, Oxford University. http://www.fhi.ox.ac.uk/wp-content/uploads/brain-emulation-roadmap-report1.pdf. Accessed 29 June 2013

91. Williams GC (1999) The Tithonus error in modern gerontology. Q Rev Biol 74(4):405–415

92. Hurst SA, Mauron A (2003) Assisted suicide and euthanasia in Switzerland: allowing a role for non-physicians. Br Med J 326:271–273

93. BBC (2013) Terry Pratchett: choosing to die. http://www.bbc.co.uk/programmes/b0120dxp. Accessed 17 Sept 2013
94. DESA (2010) World population aging 2009, United Nations. http://www.un.org/esa/population/publications/WPA2009/WPA2009-report.pdf. Accessed 27 June 2013
95. Bongaarts J (2009) Human population growth and the demographic transition. Phil Trans R Soc B 364:2985–2990
96. Lewis MJ (2007) Medicine and care of the dying: a modern history. Oxford University Press, Oxford
97. Connell PP, Hellman S (2009) Advances in radiotherapy and implications for the next century: a historical perspective. Cancer Res 69:383–392
98. Devita VT Jr, Chu E (2008) A history of cancer chemotherapy. Cancer Res 68:8643–8653
99. Hare R (1982) New light on the history of penicillin. Med Hist 26:1–24
100. Nesse RM (2000) Is depression an adaption? Arch Gen Psychiatry 57(1):14–20
101. Petronis A (2004) The origin of schizophrenia: genetic thesis, epigenetic antithesis, and resolving synthesis. Biol Psychiatry 55:965–970
102. Laćan G, De Salles AA, Gorgulho AA, Krahl SE, Frighetto L, Behnke EJ, Melega WP (2008) Modulation of food intake following deep brain stimulation of the ventromedial hypothalamus in the vervet monkey. Laboratory investigation. J Neurosurg 108(2):336–342
103. Henderson MB, Green AI, Bradford PS, Chau DT, Roberts DW, Leiter JC (2010) Deep brain stimulation of the nucleus accumbens reduces alcohol intake in alcohol-preferring rats. Neurosurg Focus 29(2):E12
104. Burgess A (1962) A clockwork orange. William Heinemann, London
105. Koestler A (1967) The ghost in the machine. Hutchinson, London
106. Hulkower R (2011) From sacrilege to privilege: the tale of body procurement for anatomical dissection in the United States. Einstein J Biol Med 27(1):23–26
107. Shildrick M, McKeever P, Abbey S, Poole J, Ross H (2009) Troubling dimensions of heart transplantation. Med Humanit 35:35–38
108. Sanner MA (2003) Transplant recipients' conceptions of three key phenomena in transplantation: the organ donation, the organ donor and the organ transplant. Clin Transplant 17(4):391–400
109. Kemph JP (1966) Renal failure, artificial kidney and kidney transplant. Am J Psychiatry 122:1270–1274
110. Young EJ, Aceti M, Griggs EM, Fuchs RA, Zigmond Z, Rumbaugh G, Miller CA (2013) Selective, retrieval-independent disruption of methamphetamine-associated memory by actin depolymerization. Biol Psychiatry. doi:10.1016/j.biopsych.2013.07.036
111. Shirow M (1995) Ghost in the shell. Dark Horse Comics, Milwaukie
112. Lerner J, Wulf J (2007) Innovation and incentives: evidence from corporate R&D. Rev Econ Stat 89:634–644
113. Agin D (2007) Junk science: an overdue indictment of government, industry, and faith groups that twist science for their own gain. St Martin's Press, New York
114. Schön JH, Meng H, Bao Z (2001) Self-assembling monolayer organic field-effect transistors. Nature 413:713–716
115. Beasley MR, Kroemer H, Kogelnik H, Monroe D, Datta S (2002) Report of the investigation committee on the possibility of scientific misconduct in the work of Hendrik Schön and coauthors. http://publish.aps.org/reports/lucentrep.pdf. Accessed 27 June 2013
116. Schön JH (2003) Retraction. Nature 422(6927):92–93
117. Schön JH (2002) Retraction. Science 298(5595):961
118. Reich ES (2010) Plastic fantastic: how the biggest fraud in physics shook the scientific world. Palgrave Macmillan, Basingstoke, Hampshire, UK
119. Van Noorden R (2011) Science publishing: the trouble with retractions. Nature 478(7367): 26–28
120. Borzelleca JF (2000) Paracelsus: herald of modern toxicology. Toxicol Sci 53(1):2–4
121. Oberdörster G, Oberdörster E, Oberdörster J (2005) Nanotoxicology: an emerging discipline evolving from studies of ultrafine particles. Environ Health Perspect 113(7):823–839
122. FDA (2012a) Guidance for industry: assessing the effects of significant manufacturing process changes, including emerging technologies, on the safety and regulatory status of food

ingredients and food contact substances, including food ingredients that are color additives. http://www.fda.gov/downloads/Cosmetics/GuidanceComplianceRegulatoryInformation/ GuidanceDocuments/UCM300927.pdf. Accessed 18 Sept 2013

123. FDA (2012b) Guidance for industry: safety of nanomaterials in cosmetic products. http:// www.fda.gov/downloads/Cosmetics/GuidanceComplianceRegulatoryInformation/ GuidanceDocuments/UCM300932.pdf. Accessed 18 Sept 2013

124. Tyner K, Sadrieh N (2011) Considerations when submitting nanotherapeutics to FDA/CDER for regulatory review. In: McNeil SE (ed) Characterization of nanoparticles intended for drug delivery. Humana Press, New York

125. Hamburg MA (2012) Science and regulation. FDA's approach to regulation of products of nanotechnology. Science 336(6079):299–300

126. REACH (2008) Regulatory aspects of nanomaterials 2008. http://ec.europa.eu/nanotechnology/pdf/comm_2008_0366_en.pdf. Accessed 29 June 2013

127. SCENIHR (2010) Scientific basis for the definition of the term "Nanomaterial". http:// ec.europa.eu/health/scientific_committees/emerging/docs/scenihr_o_030.pdf. Accessed 29 June 2013

128. Brouwer N, Weda M, Van Riet-Nales D, De Kaste D (2010) Nanopharmaceuticals: implications for the European pharmacopoeia. Pharmeuropa 22:5–7

129. EC (2013a) Consultation on the modification of the REACH Annexes on Nanomaterials. http://ec.europa.eu/environment/consultations/nanomaterials_2013_en.htm. Accessed 18 Sept 2013

130. EC (2013b) Nanotechnology. http://ec.europa.eu/nanotechnology/index_en.html. Accessed 18 Sept 2013

131. Nowack B, Krug HF, Height M (2011) 120 years of nanosilver history: implications for policy makers. Environ Sci Technol 45:1177–1183

132. Braydich-Stolle L, Hussain S, Schlager J, Hofmann MC (2005) In vitro cytotoxicity of nanoparticles in mammalian germline stem cells. Toxicol Sci 88(2):412–419

133. EFSA (2008) Inability to assess the safety of a silver hydrosol added for nutritional purposes as a source of silver in food supplements and the bioavailability of silver from this source based on the supporting dossier. EFSA J http://www.efsa.europa.eu/en/scdocs/doc/ans_ ej884_Silver_Hydrosol_st_en.pdf?ssbinary=true. Accessed 29 June 2013

134. EC (2009) Commission Regulation (EC) No 1170/2009. Official Journal of the European Union. http://eur-lex.europa.eu/LexUriServ/LexUriServ.do?uri=OJ:L:2009:314:0036:0042: EN:PDF. Accessed 18 Sept 2013

Chapter 14
Nanomedicine as a Business Venture

Olivier Fontaine, Bojan Boskovic, and Yi Ge

List of Abbreviations

BIO	Biotechnology Industry Organisation
CAGR	Compound Annual Growth Rate
EC	European Commission
EPO	European Patent Office
ESF	European Science Foundation
ETPN	European Technology Platform on Nanomedicine
EU	European Union
FDA	Food and Drug Administration
FP	Framework Programme
GCP	Good Clinical Practice
GDP	Gross Domestic Product
GMO	Good Manufacturing Organisation
GMP	Good Manufacturing Practice

O. Fontaine (✉)
European Technology Platform on Nanomedicine (ETPN),
Steinplatz 1, 10623, Berlin, Germany
e-mail: olivier.fontaine@nanobiotix.com

B. Boskovic
Cambridge Nanomaterials Technology Ltd,
14 Orchard Way, Lower Cambourne, CB23 5BN Cambridge, UK

Y. Ge (✉)
Centre for Biomedical Engineering, Cranfield University,
Cranfield, Bedfordshire, MK43 0AL, UK
e-mail: y.ge@cranfield.ac.uk

© Springer Science+Business Media New York 2014
Y. Ge et al. (eds.), *Nanomedicine*, Nanostructure Science and Technology,
DOI 10.1007/978-1-4614-2140-5_14

IP   Intellectual Property
IPTS   Institute for Prospective Technological Studies
JRC   Joint Research Centre
NDDS   New Drug Delivery Systems
NIH   National Institute for Health
R&D   Research and Development
SME   Small and Medium Enterprise
VC   Venture Capital
WoS   Web of Science

14.1 The Current Nanomedicine Market

In the last few years, many marketing and/or scientific reports on nanomedicine have been published, based on the 3–5 years of data accumulation. Among them, the most highlighted reports in Europe include: (1) Scientific forward look on nanomedicine [1]; (2) Strategic research agenda for nanomedicine [2]; (3) NanoMed round table extended report [3]; and (4) Contribution of Nanomedicine to Horizon 2020 [4].

There is another important report [5] which executive summary was published first in Nature Biotechnology in 2006 [6], in addition to a comprehensive review on nanobiotechnology in the medical sector [7] and a very recent paper [8] on the state of investigational and approved nanomedicine products. Unfortunately, up-to-date data on the market landscape of nanomedicine still could be limitedly found on a free access basis due to the exceptionally rapid development of nanomedicine, intellectual property issue and business information/data protection.

14.1.1 Nanomedicine: A Global Market Analysis

14.1.1.1 A Market Overview

In 2004, there were already 38 nanotechnology-enhanced medical products on the market with estimated total sales of EUR 5.4 billion [5]. According to an European Science Foundation (ESF) Report published in 2005 [1], drug delivery applications account for three-quarters of the total nanomedicine market, with a special emphasis on novel drug delivery systems (NDDS) (23 products on the market). This statement was also reinforced by a French report stating that drug delivery applications are accounted for 58 % of the 36 nanoproducts already on the market (Bionest Partners and LEEM [9]). The reasons of such commercial interest and success will be addressed and discussed later in this chapter.

As further described in Wagner and his co-authors' report, for drug delivery, most products are based on liposomal or virosomal formulations [5]. In vivo imaging is highly represented by iron nanoparticles for liver tumours, whereas colloidal gold is mostly used for in vitro *diagnostics* and lateral flow tests. Biomaterials on the market are for dental filling and repair or to tackle bone defects, while active implants are to cope with heart failure.

In the field of therapeutics, anticancer drugs seems represent the largest product segment of the nanomedicine market. According to the FP7 EuroNanoMed's Strategic Agenda [10], "of the 65 nanomedicine related trials identified in the. ClinicalTrials. gov registry, 62 were related to cancer treatment". Furthermore, as presented in a market report [11], cancer therapy will continue to be a top priority until 2016.

In terms of commercial nano-products, liposomes have one of the longest development histories, with successful drugs already on the market, such as Ambisome, a liposomal injection against fungi infection produced by Gilead (CA, USA) c.

14.1.1.2 A Breakdown per Types of Companies

207 companies based on nanomedicine activities were reported by the ESF in 2005, with more than 150 startups and small and medium enterprises (SMEs) focusing on nanomedicine R&D projects [5]. Furthermore, for the product pipeline, the nano-medicine related companies seems focus on drug delivery applications. In fact, 56 % of companies are involved in developing new drug delivery systems (NDDS), for 80 % of the nanomedicine market share. However these products represent small market values and focus on diseases of small patient groups, consequently representing niche markets [5]. It is worth noticing that although 16 % of nanomedicine companies are related to in vitro diagnostics, the outcome on the product pipeline is only of 6 %. Overall, 46 % of nanomedicine products are developed or co-developed by US companies, against 37 % for EU companies.

Among successful companies in Europe, there are three companies that are at the centre of attention and were explicitly presented during the EU FP7-funded NanoMed Round Table Extended Report [3]:

- **Nanobiotix:** develops novel therapeutics based on multifunctional nanoparticles and especially to treat cancer
- **Magforce:** proposes a revolutionary way to treat tumours via the heating of nanoparticles by an external magnetic field
- **Sonodrugs:** aims at developing novel drug delivery systems via for example a triggered release of drug by focused ultrasound induced pressure or temperature stimuli. Such NDDS is aimed to treat more efficiently cardiovascular diseases or cancer.

14.1.1.3 Predicting the Nanomedicine Market

In fact, it could be hard to economically evaluate the added value of nanotechnology in a medical product [6, 12–15]. Moreover, there is still not a proper definition of *nanomedicine* and the boundaries to depict nanotechnology-enabled medical products can vary substantially from one report to another. For some sectors however,

Table 14.1 Statements and predictions on the overall nanomedicine market using two types of analysis

Market	Type of analysis	
	Consideration of the total sales	Consideration of nano-specific sales
Nanomedicine market	The global nanomedicine sector will grow from $53 billion in 2009 to more than $100 billion in 2014 [27]	The nanomedicine market will grow to around $12 billion in 2012 [5]
	It actually reached $72.8 billion in 2011 and is expected to grow to $130.9 billion in 2016 [11]	The healthcare nanomarket will grow from $6.8 billion in 2007 to almost $29 billion in 2014 [18]
	The world nanomedicine market will cross $160 billion by 2015 [29]	
Nanobiotechnology market	The National Science Foundation projected the nanobiotechnology-based market to reach $300 billion by 2016 in the US alone and the molecular imaging market to reach $45 billion in 2010 [16, 17]	The Biotechnology Industry Organization (BIO) projected the biotech revenues to grow to $90 billion in 2008 [16, 17]

assumptions can be made and lead to a more specific market analysis. For example, the Institute for Prospective Technological Studies (IPTS) states that: "on average 30 % of the value of a drug is added by a NDDS" [5]. As a general trend, two types of predictions could be highlighted when comparing different scientific papers and marketing reports: (1) some analysts use the total sales to depict the nanomedicine market and its predictions; (2) some others try to estimate more specifically the nano-related commercial outputs of healthcare products.

The following table (Table 14.1) demonstrates the difference made by using two types of prediction analysis:

Despite the divergence in different resources, all these statements highlight how important the nanomedicine market is in terms of revenues, sales and commercial outcomes, and how fast it is growing. Whether analysts refer to the nanomedicine market, the nanobiotechnology market, or one of its leading applications such as anti-cancer products, the compound annual growth rate (CAGR) always exceeds 10 %.

Some other reports, such as IPTS's report [5] and BBC Research LCC's report [11], have emphasized on the predominant role that nanomedicine would own in healthcare applications and its impact on the worldwide economy.

14.1.2 Nanomedicine: An Overview of the Research and Development Landscape

Assessing the publications' activity (i.e. the number of published scientific papers) of a country or organization could be one of the most efficient ways to obtain a good representation of its research trends and perspectives. Important databases are at disposal to search for publications by relevant keywords. One of the most widely used and comprehensive database is the Web of Science (WoS) database maintained by Thomson Reuters.

By exploring the WoS database, it was found that in 2004 nanomedicine represented 4 % of the nanotechnology research with about 1,400 scientific publications out of 34,300 for nanotechnology. The United States is the leading country in the field of nanomedicine research, accounting for 32 % of the publications. EU follows, with Germany being the most active country with 8 % of the worldwide publications. As for the nanomedicine products' distribution, drug delivery is the top research area, accounting for 76 % of the scientific papers, followed by in vitro diagnostics (11 %) and biomaterials (6 %) [6]. USA is also the only country in 2004 to have its number of patents in the field exceeding the one of publications, showing that nanomedicine is still at an early phase of development with limited commercial outcomes yet [5].

Nanomedicine also receives growing shares of public funding and strategic initiatives supplied and developed by some leading economic bodies/countries, such as USA, EU, Japan and China, to promote the research of nanotechnology. Unfortunately, there are no data available on public funding of nanomedicine research merely. However, the share of nanotechnology publications focusing on healthcare could be used as an indicator in such case. Hence, it is assumed that "in the EU25 about 5 % of the nanotechnology funding is spent on medicine-related research" and about 20 % on life sciences more generally [5].

14.1.3   Nanomedicine: An Overview of the Patent Landscape

14.1.3.1   A New Era of Commercial Development

Studying the patent landscape via official patent database such as the European Patent Office (EPO) could form an efficient way to obtain an overall picture of the development of a novel technology. Since patents are a central milestone to entering the market and a key step towards commercialization, such analysis can provide some significant clues on the current trends and future perspectives of the commercial outcomes and the maturity of the market. The patenting activity thus could be an intermediate step in the supply chain of nanomedicine, following research and publications and preceding commercial output and on-the-market products.

For nanomedicine itself, a symbolic line have already been crossed: since the beginning of the millennium, the trend of patenting activity is now approximately rising to a synchronous pace with the pace of publications [16, 17]. The number of filed patents grew from 2160 in 1989 to 7763 in 2002. According to Wagner and his colleagues [6], the related patenting activities have even outpaced research publications since 2001. 2000 patents were filed in the field of nanomedicine sector in 2003, compared to 220 in 1993.

It seems that nanomedicine has entered a new era of commercial development and started to fulfil its initial hype. The increase of commercial outputs of research findings is notably attributed to the increased amount of research investment from government, corporate and private sources, along with numerous initiatives to develop translational nanomedicine.

14.1.3.2 Sectorial and Country Breakdown of Patents in Nanomedicine Worldwide

In the field of nanomedicine, the dominant patent sector is drug delivery with a share of 59 % in 2006, followed by in vitro diagnostics (14 %), Imaging (13 %) and others (e.g. drugs) (14 %) [6].

USA was the leading country for the nanomedicine patents, with a share of 54 % worldwide in 2006. In comparison, Europe held only 25 % share despite the fact that it was the world leader for the publication of nanomedicine papers. Herein, Germany itself had an outstanding activity with a share of 12 %. Meanwhile, Asia held 12 % share [5].

According to another report from Scrip Insights [18], however, Europe gained a substantial increase to 36 % of the overall number of relating patents in 2009, due to the increasing R&D efforts and the greater market participation of European healthcare players.

14.2 The Nanomedicine Business Environment

14.2.1 Introduction: Key Players in the Nanomedicine Business Environment

Nanotechnology is at the interplay between governments, regulatory bodies, start-ups, venture capitalists, start-ups and entrepreneurs [19]. Consequently, nanomedicine could expect a similar environment. However, the environment becomes much more complex when it comes to medical applications and healthcare drivers since (1) public perception and ethical issues have now become crucial to be taken into consideration; (2) pharmaceutical companies as well as clinical trial centres now have entered the dynamic environment as major players in the development of nano-related medical products; and (3) regulation bodies are now at the centre of marketing approval, making impacts on each stakeholder. A re-evaluation of the nanomedicine business environment is thus required.

The key players and their impact on the nanomedicine business environment are given in Table 14.2.

As highlighted in Table 14.2, pharmaceutical companies play a crucial role in the current nanomedicine supply chain. Since they are the only industry structure to provide enough funding for the development costs in clinical trials (accounting for millions of dollars), they are at the core of the business process. Moreover, the uptake by pharmaceutical companies is also very beneficial to start-ups and SMEs:

- The licenses will provide an additional revenue stream, whereas partnerships will offer them the distribution network and structure of the pharmaceutical industry;
- The uptake of on-going products after proof of concept can be seen as a liquidity exit in a much shorter timeline, which could greatly encourage and attract investors to fund start-ups and SMEs based on this business model.

Table 14.2 Key players and their impact on the nanomedicine business environment [19]

Key stakeholder in the nanomedicine supply chain	Impact on the supply chain
Academia (Research)	Innovation driver
	Availability of technology and equipment
Start-ups and SMEs	Risks are assumed in this most upstream step of the business process
	Innovation drivers
	Suppliers of nanomaterials for nano-based devices, drugs and platforms
Pharmaceutical companies	Uptake of on-going products after proof of concept: developing costs in clinical trials are so important that it is hard for SMEs to do without pharmaceutical companies
	Their distribution networks is crucial
Medical and clinical trials centres	Essential to ensure Good Clinical Practice (GCP) and to maximize the chance of marketing approval
Governments	Their policy can impact innovation by attracting companies, minds and projects; and impact a few steps downstream by promoting efficient structures to conduct research ideas into manufactured products
Regulatory bodies	Agencies such as the FDA are responsible for marketing approvals of each novel medical product
	Specific legislation can facilitate technology or IP transfer and promote research
Funding sources/investors	Without capital, no research can be carried out and no innovation can possibly be translated to effective products and medical applications
	Start-ups and SMEs cannot afford by themselves developing costs along with infrastructure investments and IP strategy
Supportive infrastructures (consortia, government initiatives, research and development platforms…)	Essential to promote research, increase public awareness and therefore consumers' dynamics
	Important player in bridging the gap between industry and academia for resources gathering and more efficient translational nanomedicine

The necessity to rely on pharmaceutical companies could therefore explain why NDDS are currently highly dominant in research activities, patents filings and commercial products of nanomedicine.

If pharmaceutical companies are the main resource to monetize intellectual property, governance and venture capitalists could be another two important sources in terms of funding bodies/providers for companies which are willing to undertake risky R&D in nanomedicine. Between these two sources, venture capitalists seems more suitable to invest in nanomedicine-related start-ups since they could not only provide a source of funding, but also offer a "governance structure" [20] and make additional contributions by sharing their expertise in market analysis and management.

14.2.2 Nanomedicine Business Drivers

The study of business drivers delivers another complementary view on the business environment. Drivers could result more or less directly from key players (such as the availability of capital results from funding bodies and venture capitalists) and impact on the dynamics of the business.

Business drivers can be defined loosely as the main factors and resources which provide the essential marketing, sales, and operational functions of a business, are of paramount importance. For nanomedicine business drivers, they should particularly encompass all factors to drive commercialization in nanomedicine.

In a general way, the research and development strategic agenda of nanomedicine could result from two driving forces: (1) technological innovation and science progress (upstream technological push); and (2) healthcare needs (downstream clinical pull). The technological push may lead to products and applications resulting from science/technology breakthrough and will answer "what can be done with such progress?", whereas the clinical pull is able to tailor products to healthcare needs and will answer "how can this urgent need be technologically addressed?".

If the research in nanomedicine is highly pushed by technology, governance and stakeholders would then like to see a shift to a more clinically driven development, to ensure optimal healthcare benefits and maximal acceptance by the public, patients and clinicians [7]. However, technological drivers are still crucial, since genomics and proteomics lead a better understand and redefinition of diseases while nanoscale manipulation of proteins and DNA is necessary for enhancing diagnosis and treatments.

There is a need to emphasize that nanomedicine encompasses a highly diversified range of products and technologies, which fall in different markets (size or maturity), industries and regulations. Hence, specific drivers and dynamics would have to be considered for each sector of application, for each type and size of companies, and in each country's regulatory landscape, which is beyond the scope of this study. The following Table (Table 14.3) is a non- exhaustive list of influencing factors for business and commercialization of nanomedicine together with an estimated assessment in terms of their roles as drivers or hurdles based on a self-study in the related fields and some published resources [4, 10, 16, 17, 19, 21–27]. For the intellectual property (IP) landscape, current issues and strategies are focusing on patents as they are most attractive to investors due to the market exclusivity and the high guarantees and protections they offer. However, for IP strategy it can further involve trademarks, trade secrets, and copyrights [19].

14.2.3 Business Models and Strategies

As described earlier, the current nanomedicine research and development is driven by start-ups and SMEs. According to the IPTS Report [5], these innovative companies rely mainly on three types of business models, depending on the category of their novel products (Table 14.4).

Table 14.3 A non-exhaustive list of influencing factors for business and commercialization of nanomedicine together with an estimated assessment in terms of their roles as drivers or hurdles

Influencing factors for business and commercialization of nanomedicine	Roles as drivers or hurdles
IP landscape and strategy	Driver — Hurdle
Healthcare costs	Driver — Hurdle
Aging population	Driver — Hurdle
Availability of funding and venture capital especially	Driver — Hurdle
Structure for translational nanomedicine	Driver — Hurdle
Regulatory landscape	Driver — Hurdle
Public perception and acceptance	Driver — Hurdle
Manufacturing considerations	Driver — Hurdle
Technology status promises	Driver — Hurdle

The crucial role played by pharmaceutical companies in the current development of most nanomedicine-related products is well conveyed in the business models pursued by start-ups and SMEs.

In terms of business strategies, the market players have four main types of strategic initiatives at their disposal to increase revenue and boost market growth in the competitive and fragmented landscape of nanomedicine [18]:

- Investment in R&D activities to upgrade their existing technologies for newer applications or to develop new products (58 % of major players' strategy for the 2007–2009 period);
- Collaboration and agreements with research institutions to explore avenues for new products and applications (25 %);
- Launching of new products (12 %);
- Mergers and acquisitions (5 %);

Table 14.4 Business models of nanomedicine start-ups and SMEs (Modified from [5])

Business model Category of product	Description	Start-ups and SMEs Examples
Development of nano-pharmaceuticals or medical devices	The aim is to develop a proprietary product pipeline by bringing novel medical devices, drugs, or conventional drugs with NDDS to market.	Inex, USA
	The company goes through the development until proof of concept is acquired. Afterwards the partnerships with pharmaceutical companies often take the product through clinical trials and offer the required distribution networks	Starpharma, Australia Idea, Germany
Development of nano-platforms	The aim is to provide added value to second-party products. For drug delivery companies, they usually focus on a particular delivery technology.	Pharmasol, Germany
	The technology is then often licensed to pharmaceutical companies, as it may have been directly customised to their needs.	Eiffel Technologies, Australia
Development and manufacturing of high-value nanomaterials	The aim is to manufacture and provide specific nanomaterials (such as carbon nanotubes or quantum dots) that will be used in nano-enhanced drugs or medical devices.	Biogate, Germany (silver nanoparticles)
	Such nanomaterials are often directed to a specific area of application such as implants, sensors and lateral flow tests.	British Biocell, UK (gold nanoparticles for lateral flow tests) Raymor Industries, USA (nanosized titanium powders for implants)

As a result of the early stage of nanomedicine's development, R&D activities highly dominant, accounting for 58 % of the business strategies. Whatever the initiative, drug formulation and delivery accounts for the main application, with 77 % of the collaborations and 58 % of the R&D and new products launches [18].

14.2.4   The Nanomedicine Business Environment: An Overview of Its Dynamics

A summary chart is given in Fig. 14.1, illustrating the overview of the nanomedicine business environment and its dynamics based on the earlier discussions.

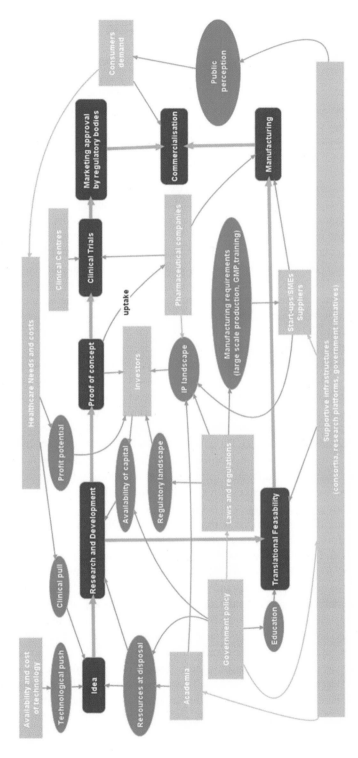

Fig. 14.1 The nanomedicine business environment. *Blue block*: nanomedicine supply chain; *orange block*: business players; *green block*: business drivers; *arrows*: impact

14.2.5   Current Issues and Future Perspectives

14.2.5.1   Current Issues and Strategic Recommendations

At the moment, most of the emerging fields of research and development are often driven by government-funded projects or initiatives, since the venture capital (VC) capitalists usually consider too much about the maturity degree of science/technology and the uncertainty of business environment. According to Jackson and his colleagues' study [19], attractiveness of start-up companies to VCs is embodied by very short time-to-market horizon (less than 2 years), an established customer base, and a successful management structure with experienced executives. Unfortunately, the current development status of nanomedicine does not quite meet those characteristics and conditions. The development of a nanomedicine product could takes 10–15 years and the public perception is highly unstable and impressionable by the toxicological concerns. Furthermore, no golden standard is yet established in the commercial steps towards business success or in the managing structure and expertise to gather. The evolving and uncertain regulatory landscape along with the emerging "patent's thickets" [16] doesn't lighten this picture of the business environment. Finally, VC often enters the funding stream after the proof of concept is established. Thus VC is unable to not represent the main funding body when investments are needed to launch an innovative yet risky start-up.

Although the availability of venture capital seems currently one of the weak parts of the European market, as one of the emerging and most promising fields worldwide, nanomedicine has benefited from a wide diversity of funding resources and ongoing government initiatives to promote its sustainable development. As a result, some highly ambitious programs have already arisen, with the creation of major clusters and centres of excellence for gathering resources and people around nanomedicine, to cope with its multidisciplinary and highly complex dimensions. In this context, governments and supportive infrastructures have initiated a series of initiatives, workshops and round tables to carefully frame the future development and commercialization environment of nanomedicine.

For example, the European Technology Platform on Nanomedicine (ETPN), a supportive infrastructure funded by FP7, established its first strategic research agenda in 2006. In its comprehensive report [2], some general recommendations were provided: focus should be made around increased innovation and reduced toxicity in order to reassure investors and promote a less risky and uncertain business environment; delays in commercialization should be addressed, with adequate and specific manufacturing standards and less complex and stringent regulatory systems; increased collaboration, consultation and cooperation will be crucial as well to avoid fragmentation and lack of coordination in this emerging technology.

In terms of strategic considerations, the following criteria, along with issues to be addressed, were further identified by ETPN (see Table 14.5) [2] as areas of priority to provide nanomedicine with a sustainable developing environment:

There are some other important reports and collaborative initiatives include: (1) Scientific forward look on nanomedicine [1]; (2) NanoMed round table extended

Table 14.5 Strategic considerations and issues to be addressed for a sustainable development of nanomedicine [2]

Strategic consideration	Issues to be addressed
Public acceptance	Transparent dialogue to avoid overflow of negative opinion due to toxicological concern
	Need to speak from the perspective of nanomedicine, not nanotechnology
Risk assessment	Specific research to answer safety-related basic science questions: in vitro models, identification of parameters, and high throughput screening
Regulatory framework	Address adequacy and appropriateness of current system in terms of novel products and not novel enabling technologies
IP rights	New model is to be investigated by ETPN
Required Research Infrastructure	Increase proximity between experts and facilities of different areas by larger clusters and centres of excellence
	Increase translation of research results to clinic of patients
Education and training to facilitate adoption of technology in hospitals	Need to develop and synchronize regional education schemes
	Develop training of industry and clinical personnel at all levels
Manufacturing costs	Commercialized products should have high potency and targeting in order to counter-balance with the complexity of manufacturing

report [3]; and (3) Concept for a European Infrastructure in Nanobiotechnology [25]. More recently, ETPN published a report on Contribution of Nanomedicine to Horizon 2020 [4], which is a White paper to the Horizon 2020 Framework Programme for Research and Innovation. It gave some specific recommendations from the nanomedicine community in order to exploit nanomedicine's full potential and for products to be brought efficiently to the market:

- A specific funding programme should be established for translational nanomedicine topics in order to refocus projects and adjust funding allocation criteria.
- A novel translational infrastructure should be established by federating research centres and clinical centres to de-risk innovation and leverage capabilities.
- Translational know-how should be recognised and more easily accessible to encourage a shift from disruptive and fundamental research to more translation focused projects.

14.3   Conclusions and Future Perspectives

In conclusion, nanomedicine is often seen as the area with the most promising and visible applications of nanotechnology on the short and long term. However, it is still in an early stage of commercialisation and its environment is currently undergoing high scrutiny as products are entering the market. At the moment, drug delivery, imaging and in-vitro diagnostics remain the key applications in terms of recent dynamics, market attractiveness, research interests and level of technology readiness.

Early in 2009, Wiek et al. highlighted that the interplay between very similar context variables (business players and drivers) in nanotechnology could lead to highly different market situations [28]. For the development and commercial outcomes of nanomedicine, there could be three hurdles and threats:

- Health and environmental toxicological effects proven by long-term risk assessments.
- Media influence
- Risk-averse public awareness

The on-going IP landscape may also discourage large and pharmaceutical companies of structural changes to embrace nanomedicine research and development. In addition to different market access to companies, unforeseen future obstacles and highly impacting factors might also come from a major and emerging societal issue: the access of treatments to patients. Costs will substantially vary from one hospital to another due to different budget envelopes per disease, and from one country to another due to different reimbursement policies [3].

Nonetheless, it has become a more and more solid fact that the nanomedicine development is currently witnessing a shift in stakeholders' wills and awareness in the priority issues to address and in the necessary environment to provide. In the near future, deeper understanding of the scientific and technological processes coupled with enhanced analytical tools for early characterization of toxicological profiles and a more established business model for efficient commercialization of nanomedicine products, will be able to provide this emerging "sustainable governance" with all the required knowledge for evidence-based decisions and for promoting the most beneficial applications for our society.

Heavily increasing funding programs and infrastructures, proactive initiatives and collaboration, as well as great efforts toward appropriate regulation and societal formation, have laid the foundation and framed a very bright future for nanomedicine. The next coming years will be crucial in building the path towards the future socio-economic impact of the nanomedicine era.

References

1. European Science Foundation (2005) Forward look on nanomedicine. http://www.nanophar-maceuticals.org/files/nanomedicine.pdf. Accessed 18 July 2013
2. European Technology Platform on Nanomedicine (2006) Strategic research agenda for nanomedicine. ftp://ftp.cordis.europa.eu/pub/nanotechnology/docs/nanomedicine_bat_en.pdf. Accessed 18 July 2013
3. Institut für Philosophie (2009) NanoMed round table extended report. http://www.philosophie.tu-darmstadt.de/media/institut_fuer_philosophie/diesunddas/nordmann/nanomed.pdf. Accessed 18 Dec 2013
4. European Technology Platform on Nanomedicine (2013) Contribution of nanomedicine to horizon 2020. http://www.etp-nanomedicine.eu/public/press-documents/publications/etpn-publications/etpn-white-paper-H2020. Accessed 18 Jan 2014
5. Wagner V, Hüsin B, Gaisser S, Bock AK (2008) Nanomedicine: drivers for development and possible impacts. http://ipts.jrc.ec.europa.eu/publications/pub.cfm?id=1745. Accessed 18 Jan 2014

6. Wagner V, Dullaart A, Bock A, Zweck A (2006) The emerging nanomedicine landscape. Nat Biotechnol 24(10):1211–1217
7. Hüsing B, Gaisser S (2006) Nanobiotechnology in the medical sector – drivers for development and possible impacts. http://www.isi.fraunhofer.de/isi-media/docs/t/de/publikationen/ISI_Nanobiomedicine_WP3_revised-290320061.pdf. Accessed 18 July 2013
8. Etheridge ML, Campbell SA, Erdman AG, Haynes CL, Wolf SM, McCullough J (2013) The big picture on nanomedicine: the state of investigational and approved nanomedicine products. Nanomedicine 9(1):1–14
9. Bionest Partners and LEEM (2006) Nanomedicine study. http://www.leem.org/sites/default/files/1425.pdf. Accessed 18 July 2013
10. EuroNanoMed (2009) Strategic agenda for EuroNanoMed. http://www.euronanomed.net/files/Strategic_Agenda_for_EuroNanoMed.pdf. Accessed 18 July 2013
11. BCC Research LLC (2012) Nanotechnology in medical applications: the global market. http://www.bccresearch.com/market-research/healthcare/nanotechnology-medical-applications-global-market-hlc069b.html. Accessed 18 July 2013
12. Hafner A, Lovrić J, Lakoš G, Pepić I (2014) Nanotherapeutics in the EU: an overview on current state and future directions. Int J Nanomed 9:1005–1023
13. Harris E (2014) Industry update: the latest developments in therapeutic delivery. Ther Deliv 5(4):381–386
14. Lévy L (2014) Europe as leaders in nanomedicine: let's go for it! Nanomedicine 9(4):389–391
15. Menaa F (2014) Global financial model for responsible research and development of the fast growing nanotechnology business. J Bus Financ Affairs 3(1):1–2
16. Bawa R (2005) Will the nanomedicine "patent land grab" thwart commercialization? Nanomedicine 1(4):346–350
17. Bawa R, Bawa SR, Maebius S, Flynn T, Wei C (2005) Protecting new ideas and inventions in nanomedicine with patents. Nanomedicine 1(2):150–158
18. Scrip Insights (2010) Nanotechnology in Healthcare: Market outlook for applications, tools and materials, and 40 company profiles. http://www.reportlinker.com/p0180553-summary/Nanotechnology-in-Healthcare-Market-outlook-for-applications-tools-and-materials-and-40-company-profiles.html. Accessed 18 July 2013
19. Jackson MJ, Whitt MD, Handy RG, Robinson GM, Whitfield MD (2009) Commercialization of nanotechnologies: technology transfer from University Research Laboratories. In: Ahmed W, Jackson MJ (eds) Emerging nanotechnologies for manufacturing, 1st edn. Elsevier, Oxford, pp 251–260
20. Pisano GP (2010) The evolution of science-based business: innovation how we innovate. Ind Corp Change 19(2):465–482
21. LuxCapital (2003) The NanoTech ReportTM 2003: Investment overview and market research for nanotechnology (Volume II). http://www.altassets.net/pdfs/nanotechreportluxcapital.pdf. Accessed 18 July 2013
22. Brower A (2005) Report portrays nanotechnology as driver of reduced R&D costs. Biotechnol Healthc 2(2):19–20
23. Bawa R (2007) Patents and nanomedicine. Nanomedicine 2(3):351–374
24. Munos B (2009) Lessons from 60 years of pharmaceutical innovation. Nat Rev Drug Discov 8:959–968
25. EuroNanoBio (2010) Concept for a European infrastructure in nanobiotechnology. http://www.nanowerk.com/nanotechnology/reports/reportpdf/report131.pdf. Accessed 18 July 2013
26. Prescott C (2010) Regenerative nanomedicines: an emerging investment prospective? J R Soc Interface 7(6):S783–S787
27. Morigi V, Tocchio A, Pellegrinelli CB, Sakamoto JH, Arnone M, Tasciotti E (2012) Nanotechnology in medicine: from inception to market domination. J Drug Deliv. doi:10.1155/2012/389485
28. Wiek A, Gasser L, Siegrist M (2009) Systemic scenarios of nanotechnology: sustainable governance of emerging technologies. Futures 41(5):284–300
29. Global Industry Analysts Inc (2009) Nanomedicine – a global market report. http://www.strategyr.com/MCP-1482.asp. Accessed 18 Jan 2014

Chapter 15
What Can Nanomedicine Learn from the Current Developments of Nanotechnology?

Sirikanya Chokaouychai, Dan Fei, and Yi Ge

15.1 Introduction

Nanotechnology is defined as the design, control, manipulation, synthesis, production and application of properties and functionalities of any structure, device, or system, which has one of its dimensions between 1 and 100 nm, by controlling its shape and size [1]. Expanding from its original definition, nanotechnology can be applied to and combined with various fields of science and gives rise to emerging sciences which greatly improves technology and, ultimately, the quality of life. An excellent example to this is the combination between nanotechnology and medicine – known as the emerging 'nanomedicine'. Therefore, nanomedicine can be defined, in general, as the application of nanotechnology to medicine to create advanced diagnostics and therapeutics for disease treatment and prevention from nanoscale, using knowledge, principles, and techniques from nanotechnology [1–4].

The starting point of nanotechnology in the human history began when Richard Feyman gave his infamous lecture in 1959 'There's Plenty of Room at the Bottom' stating the idea of manipulating individual atoms using larger equipment to produce relatively small matters. However, it was only in 1974 that the term 'nanotechnology' was first invented by Norio Tanaguchi and was accepted as an official, new scientific terminology. Nanotechnology and nanosciences has begun to grow at an incredible speed since then, and their applications started to branch out in various fields. However, proper and serious attention in nanomedicine has begun since only

S. Chokaouychai • Y. Ge (✉)
Centre for Biomedical Engineering, Cranfield University,
Cranfield, Bedfordshire, MK43 0AL, UK
e-mail: y.ge@cranfield.ac.uk

D. Fei
Leicester School of Pharmacy, De Montfort University,
The Gateway, Leicester, LE1 9BH, UK

© Springer Science+Business Media New York 2014
Y. Ge et al. (eds.), *Nanomedicine*, Nanostructure Science and Technology,
DOI 10.1007/978-1-4614-2140-5_15

a few decades ago [5, 6, 7]. Table 15.1 below lists a brief summary of hallmarks relating to the evolution of nanotechnology and nanomedicine.

At nanoscale, substances exhibit unique properties and phenomena which are absent or different from when they are at macroscale [5]; based on the significance of this fact, versatile applications and implications can be made to find solutions to the related unsolved problems at macroscale. The application of nanotechnology in medicine has been brought to attention based on the fact that cellular components, activities and interactions, which are the essence and the most basic level of life, are at nanoscale (Table 15.2); hence working at the same scale might lead to solutions to current medical limitations or, at least, provide better understanding of the situation. Nanotechnology is widely applied to medical imaging, disease detection, medical analysis, drug manipulation, and modelling at nanoscale [5] to develop advanced diagnostic and therapeutic tools for curing and preventing diseases, and ultimately improve the quality of life.

Table 15.1 A brief list of nanotechnology and nanomedicine hallmarks [5]

Year	Event
1959	Richard Feyman's "There's Plenty of Room at the Bottom" lecture – the starting point of nanotechnology
1974	Establishment of the term 'nanotechnology' by Norio Tanaguchi
1979	World's first use of colloidal Au nanoparticles (NPs) in electron microscopy
1987	World's first cancer targeting using NPs coated with monoclonal antibodies
1990	World's first visualisation of atoms by the scanning tunnelling microscope (STM) invented by IBM Zurich Lab
1991	Discovery of carbon nanotube
1994	Establishment of the concept of NP-based drug delivery
1995	Liposome fabrication and usage in drug delivery
1998	Establishment of the term 'nanomedicine'

Table 15.2 Example dimensions of significant biological substances in the body [5]

Average dimensions (nm)	Substances
2,500	Human red blood cell
65–100	Exosome (vesicles from dendritic cells)
1–20	Proteins
2–4	Ribosomes
2.5	DNA (diameter)
1.2	The largest amino acid measured
0.4	A base pair in human genome
0.25	Average individual atom

From a general point of view, the scope of nanomedicine can be categorised into the following three main categories [8]:

- Medical imaging, diagnostics, and therapeutics using engineered nanoparticles and nanomaterials
- Regenerative medicine and other relevant innovative treatments
- Studies for in-depth understanding of activities, functions and mechanisms inside the body at nanoscale, and inside the cell

In this chapter, discussions about what nanomedicine can learn from the current developments in nanotechnology, and relevant topics, such as benefits, concerns, challenges and limitations, are explained. In addition, conclusions and suggestions on possible future opportunities and perspectives are stated.

15.2   From Nanotechnology to Nanomedicine

Achievement in nanotechnology opens the gate to the new era of medicine, creating opportunities for better understanding, development and invention from new perspectives. It moves medical challenges to the next step. However, everything has two sides: benefits and drawbacks. It is absolutely crucial, and it is always the aim, to minimise the adverse sides of the technology, and maximise its benefits.

Nanomedicine is a massive area, and is relatively new. Despite getting a lot of attention and funding over the past decade, the science and technology still have not been very well established and understood. Certainly, nanomedicine has been claimed to be capable of producing satisfying results and giving hopes for future medicine. However, the "negative" sides and its potential issues as well as the uncertainty have not been thoroughly studied and solved yet. As nanomedicine is closely linked to nanotechnology, it is a good idea to learn from what nanotechnology has already went through, to get a hinted starting point about what are to be marked for consideration when it comes to nanomedicine.

15.2.1   A Quick Look at the Current Development in Nanotechnology with a Critical View

Since its establishment, nanotechnology has been growing rapidly at an amazing rate. Extensive studies and experiments have been going on in various aspects of nanotechnology. Current major attentions are, for example, the development of nanoparticles and nanomaterials, nanofabrication techniques, nanoelectronics, precision engineering, nanofluidics, nanoreactors, advanced microscopy, nanometrology, nanotechnology for energy solution, nanophotonics, and nanotoxicity [9–15]. As studies and experiments go, investigators make discoveries of successful results, as well as problems and limitations. Discussion on how nanotechnologies have been successfully applied into medicine will be made in the next section, while the problems and issues of the technologies from a nanotechnology perspective will be discussed in this section.

A classic example of problems raised which could be related to nanomedicine is the property inconsistency of products from nanofabrication. In the production of nanoparticles, this problem occurs very frequently, even with the best technique available which produces the finest nanoparticles. With current technologies, nanofabrication still exhibits variation of nanoparticle product sizes in a batch [14]. Having size inconsistency implies that the entire batch of the particle will also have inconsistent property – low production effectiveness and product quality. As a real-case example, Saito et al. carried out the synthesis of single-walled carbon nanotubes (SWCNTs) via gas-phase pyrolytic method using metal nanoparticles as catalysts [16]. In their experiments, particle size inconsistency occurred during fabrication of metal nanoparticles, hence creating SWCNTs size inconsistency. Problems with inconsistency does not confine to only particle sizes, but also other important parameters, such as surface chemistry and concentration, as demonstrated in the work conducted by França et al. [17].

Another example is related to micro- and nanofluidics which are very complicated concepts and are still under investigation [18]. A micro/nanodevice inevitably requires application of micro/nanofluidics. The concepts of micro/nanofluidics are complex and challenging even to engineers as it is, in effect, a scale-down, which is the opposite to their usual work – scale-up. There are several issues that have not been understood thoroughly yet, such as interaction between fluid flow, surface forces and molecular interaction. Uniquely, at these tiny scales, surface forces become dominant. This is one of the reasons which make micro/nanofludics so different from, and much more complicated than, fluidics at the normal scale. The flow regime in these devices has unique characteristics which cannot be considered as a laminar flow, and has to be classified as a separate regime called 'microhydrodynamics'. The flow also has complicated 3-D geometries [19, 20].

A more general problem is that most of the products derived from nanotechnology currently rely on advanced technologies for their preparation and production, such as X-ray lithography for nanopattering; hence, production costs is undoubtedly expensive. Fabricating nanostructures involves several expensive technologies, for instance, as seen in the work of Liu et al. in producing palladium nanosprings [21]. Attempts have been made in trying to find alternative fabrication methods for nanotechnology related products, with the aim of reducing production costs. For example, Choi and Kim succeeded in developing an easy method to fabricate a dense nanoscale array on a large surface [22]. In addition, most of the products involving nanotechnology are still on a laboratory or pilot plant scale (i.e. still being processed in small batches), and their synthetic processes for industrial scale-up have not yet been established due to complexities and other several reasons (such as economics and profitability), which make mass production and commercialisation of these products still not practically and widely feasible.

15.2.2   A Glance of Current Advanced Nanomedicine

Before answering what nanomedicine can learn from nanotechnology so far, it is essential to understand and evaluate how nanomedicine has been making its progress. Several techniques and principles in nanotechnology, such as nanometrology, nanoparticles, nanomaterials, and nanofabrication, have been applied and implied to various medical fields. The technologies have greatly transformed medicine from its conventional practice both in diagnostics and therapeutics. It is worth noting that, several emerging medical concepts have been greatly made closer to clinical use through nanomedicine, such as regenerative medicine, theranostics, and gene therapy, as elaborated below.

15.2.2.1   Medical Diagnostics

Conventional medical diagnostic methods play important roles in medicine. However, they also have some noticeable drawbacks, such as time-consuming process and biological substance degradation. Table 15.3 shows some comparisons between conventional and nanomedical diagnostics in terms of their properties and performances. In addition to helping to modify and improve the conventional methods, nanotechnology has been applied in medical diagnostics to overcome some of their existing problems/drawbacks. Speaking in general, the major areas in nanomedical diagnostics are nanobiosensor, point-of-care (POC) medical diagnostic devices, and medical imaging.

Nanobiosensors Nanobiosensor is generally described as a biosensor at the nanoscale. A typical biosensor consists of three main parts: biological receptor element, physiochemical transducer, and detector. The first two components are critical parameters for a good biosensor. The biological receptor element should selectively and specifically binds to the desired analyte. Transduction process should be efficient so that the generated signal can be translated correctly and accurately. Hence, the design criteria for a good biosensor are selectivity, limit of

Table 15.3 Some comparisons between conventional and nanomedical diagnostics

Conventional diagnostics	Nanomedical diagnostics
Time consuming	Rapid diagnosis
Sample deterioration	Remove problems about sample deterioration
Requires a certain amount of sample to process, hence can be invasive	Requires only a very small amount of sample, hence less invasive
Difficulties from integrating parameters (resulting from various type of tests), hence requires personnel with special skills	Tends to be easy to use, hence does not require any special skill to operate
Can give inaccurate results at time, e.g. when the amount of sample is too small	Produce relatively accurate results instantly
High-cost	Low-cost

detection (LOD), response time, and signal-to-noise (S/N) ratio. Nanomaterials (e.g. nanoparticles) is one of the top candidates for the improvement of both above key components due to their unique physical and chemical properties and the ability to easily control those properties. Nanomaterials can be used to develop improved sensor coating, base, or circuit components. They can also be applied to improve the biological receptor element [23, 24].

Nanoparticles is the popular choice for studies in nanobiosensors due to several reasons, such as its unique optical property, high surface-to-volume ratio, tunable properties, high stability and biocompatibility, and non-complicated synthesis. When bound with an analyte, the overall physical and chemical property of the nanoparticle changes, thus producing detectable signal sent to the transducer. These changes, such as changes in surface plasmon resonance (SPR), electrical conductivity, or redox activity [25], could dramatically increases the sensor sensitivity. A recent example could be found from the work of Cao et al. where they reported that the plasmon shift produced from the sandwich system, having gold nanoparticles instead of secondary free antibody, was 28 times increased; hence, the signal was greatly amplified in such a manner so that it can detect the analyte at a picomolar level of concentration [26].

Several sensors can be integrated into an array called 'integrated biosensor' which is able to take different, parallel measurements simultaneously from one sample [27]. These nanobiosensors can also be combined with other nanotechnologies, such as atomic force microscopy (AFM) [28], fluorescence resonance energy transfer (FRET) [29, 30], and DNA technology [31], to improve the quality of contrast and/or add additional properties to the sensor. Recently, a novel paper-based nanobiosensor has been developed for medical diagnostic purpose by Parolo and colleagues [32].

Plastic antibody is a novel and powerful concept which greatly helps reducing several problems with biosensors. Plastic antibody is the synthetic and imprinted polymer with an affinity to bind with a specific analyte. It is produced by polymerisation of cross-linkers and functional monomer with a target molecule acting as a template. Popular templates in biological applications including proteins and small peptides. Plastic antibody has several advantages over natural antibodies. From a production point of view, it is cheaper and more stable. The product properties, such as particle size and molecular weight, can also be controlled easily during the production. For instance, [33] successfully created a nanoscale plastic antibody for the detection of a bee toxin called melittin.

Point-of-Care (POC) Medical Diagnostic Devices Compared with normal diagnostic tools, POC devices are for patients to be able to take measurement and see the results by themselves without having to visit the hospital and have their samples taken to the laboratory. Nanotechnology has made an impact on this area of medical diagnostics by offering micro/nanofluidics and nanoelectronics. POC devices tend to be of a portable size. One of the major POC principles is to take the sample from the patient as little as possible per measurement. A micro/nanofluidic system would be able to make this feasible and practical. The knowledge from nanoelectronics could greatly help with circuit and circuit component fabrication for an electronic nanoscale device. There are many researches and studies on developing 'lab-on-a-chip' devices of which mixing, separation, identification and analysis of sample fluid can be done on a small, single device [34]. Attention has been paid to the 'lab-on-a-chip' concept due to its high possibility of providing early diagnosis and therapy monitoring [35]. POC development is unfortunately not in the main stream yet, but it is a good candidate in a long term strategy [2]. Good and encouraging examples in this area include the highly-integrated lab-on-a-chip which simultaneously analyses several parameters developed by Schumacher and colleagues [36] and an attempt to create an inexpensive POC microfluidic device for viscous sample [37].

Medical Imaging   Nanoparticles, often with modified surfaces, have unique and useful properties which can be used to improve in vivo medical diagnostics by generating contrast through a selection of paths, such as radiation and magnetic field. They have been investigated and applied in various medical imaging technologies, such as magnetic resonance imaging (MRI), X-ray imaging and computer tomography (CT) to improve the efficiency of imaging tools and their contrast agents. The main purposes of the application are to make early diagnosis, track therapeutic efficiency, and obtain knowledge regarding disease development and pathology. Usually, the materials used for making contrast agents are those which are fluorescent, magnetic or paramagnetic. They can be used to localise and verify the current stage of tumour, identify the location of inflammation, verify stages of particular diseases, visualise structure of a blood vessel, and assess drug distribution and accumulation inside the body [38]. For a more advanced medical imaging, it is aimed for a method capable of detecting a single specified molecule or cell in the complex environment of human body. Since high dose of contrast agent is required for CT scan, inert materials, such as iodine-based, gold, lanthanide and tantalum nanoparticles have been chosen to make a suitable CT scan contrast agent [39]. The main focus of using nanomedical imaging technology currently is in cancer detection. Recently Chien et al. reported that administration of gold nanoparticle together with heparin produced contrast in X-ray imaging which was sufficient to see tumour microvessel (3–5 µm diameter) or extravascular diffusion [40]. For MRI contrast agents, the use of iron oxide nanoparticles is a classic example. Bae et al. developed carbon-coated iron oxide nanoparticles, improving the availability of the contrast agent [41].

15.2.2.2   Nanomedical Therapeutics

In general, the on-going research and development in nanopharmaceutics can be grouped into the following categories:

- Single, specific aspect: targeted delivery, stimuli responsive systems, controlled release, imaging, disease detection, and gene therapy
- Multifunctional nanoparticles (MFNPs): non-hybrid MFNPs and hybrid MFNPs
- Synthesis and fabrication method
- Therapeutic medical devices (e.g. cardiovascular stent)

There are further three main factors which make nanoparticles and other proper nanomaterials appealing to their application in therapeutics: very small size, unique behaviour and designable properties. Having a very small size means that they can reach sites which are previously unreachable, thereby increasing treatment effectiveness. Through several available modification techniques for nanoparticles, the particle biocompatibility, bioavailability, half-life can be increased, while the toxicity can be minimised. Apart from receiving the most attentions in cancer therapy, nanomedical therapeutic products have also been intensively investigated for providing solutions for some other major diseases, such as neurological diseases (e.g. Alzheimer's and Parkinson's diseases) cardiovascular diseases, respiratory diseases, infectious diseases (e.g. HIV and meningitis), and chronic diseases (e.g. diabetes). The field of nanomedical therapeutics generally involves nanoparticle/nanomaterial drug-delivery systems, nano-therapeutic medical devices, and special nanomedical treatments.

Nanoparticle/Nanomaterial Drug-Delivery Systems Conventionally, drugs are administered into the body and the drug molecules float around inside the body. However, not all drug molecules could efficiently arrive at the desired location: some degrade, some trapped and some cleared by the body defence mechanism. In order to cope with these problems, nanoparticles and some other nanomaterials have been widely investigated and studied. They have been successfully incorporated into the drug molecules to add or enhance properties such as biocompatibility, bioavailability, half-life, target specificity, payload, and controlled release mechanism, while minimising its toxicity effect. Coating the system with certain polymers, such as polyethylene glycol (PEG), has been shown to increase biocompatibility, bioavailability and half-life [42]. Very recently Liu et al. fabricated a complex of gold nanoshells on silica nanorattles showing increased permeability, enhanced permeability and retention (EPR) effect in tumour tissues and light conversion in vivo, while having less toxicity [43]. The concepts of antigen-antibody binding can further be applied to create target specificity by incorporating suitable ligands, such as antibody, DNA strands, and peptides, on the surface of the drug-nanoparticle complex [44]. In 2007, Hatakeyama et al. reported an anti-MT1-MMP immunoliposome complex carrying doxorubicin [45]. The complex greatly reduced tumour growth in vivo in mice. They suggested that the cellular uptake of the complex was increased due to a resulting immunoconjugation. In addition, several nanofabrication and nanoencapsulation techniques could be applied to create layers or appropriate structures so that the complex can be loaded with desired drugs of different properties, such as hydrophobicity and hydrophilicity, on a single complex, as demonstrated by Hammond [46]. In summary, the nanoparticles and other nanomaterials which have been commonly used for drug-delivery systems include [47–55]:

- Carbon nanomaterials (e.g. carbon nanotubes, fullerenes)
- Magnetic nanoparticles (e.g. iron oxide nanoparticles)
- Metal/inorganic nanoparticles (e.g. gold, silver, silica nanoparticles)
- Quantum dots
- Polymeric nanoparticles (e.g. PLGA nanoparticles)
- Solid-lipid nanoparticles

- Micelles
- Liposomes
- Dendrimers
- Multifunctional nanoparticles (MFNPs)

There are about 22 nanoparticle-content drugs which have been approved by the Food and Drug Administration (FDA) in the United States of America. Furthermore, there are about 25 nanoparticle-content drugs being investigated in clinical trials in Europe [2].

Nano-therapeutic Medical Devices   Nanotechnology has been applied not only in medical diagnostic devices, but also in therapeutic devices by improving the device's therapeutic efficiency, biocompatibility, strength or flexibilty, while minimising its adverse effects. An excellent example is the cardiovascular stent. The knowledge of nanomaterial fabrication has been used in mechanical improvement of the material used to build the stent itself [56]. Certain fabrication methods are used to improve the structure of the stent body. For example, micro-wells can be created on the surface of the metal that is used to make a stent, in order to increase drug-loading capacity of the drug eluting stent (DES). Other fabrication methods can also be used to create thin layers of polymer coating on the metal surface in order to increase the stent's biocompatibility and reduce the side effects from restenosis and thrombosis, which are side effects as a result of the interaction between the body immunological response and the stent (considered as a foreign object inside the body) [56]. There further has been a report on a successful sustained and controlled release of a DES using the layer-by-layer thin-film coating consisting of different materials for different functions [46]. Layers of drugs with different opposite charges were coated. The properties of the materials in each layer govern the kinetics of layer degradation and drug release based on the principle of mass transfer. Moreover, the technology of making a particular nanocomposite, together with surface modification, could add desired properties onto a cardiovascular stent. For example, after coating the stent with a layer of polyhedral oligomeric silsesquioxane poly (carbonate-urea) urethane (POSS-PCU), which is a nanocomposite, endothelial progenitor cell (EPC) specific antibody was successfully grafted on the composite [57]. Consequently, this device could facilitate endothelialisation of the stent to blood vessel wall, reducing problems with restenosis and thrombosis.

Innovative Medical Treatment for Specific Diseases   The unique behaviours of substance at nanoscale can be selectively used to enhance treatment efficiency. For example, there are several nanoparticles which respond to certain external stimuli. They could be applied in hyperthermia for cancer treatment. Hyperthermia literally means the condition where temperature is higher than normal. However, it can also refer to a method in cancer treatment using heat. Hyperthermia cancer treatment is a non-invasive medical treatment in which body tissue is exposed to higher temperatures to damage and kill cancer cells or to make cancer cells more sensitive to the effects of radiation and certain anti-cancer drugs. Since a too high temperature might also kill neighbouring normal cells; hence the heating process must be carefully controlled. There were problems regarding keeping the heat at the suitable level and location (problems with consistency and accuracy). For the conventional hyperthermia, heating is supplied from an external source and has to struggle through several barriers, which can be considered as resistance to heat transfer. Heat is lost during the way and the amount of heat that reaches the target is obviously less than the heat originally emitted from the heat source. However, by utilising nanoparticles in hyperthermia, it has now become feasible to achieve better consistency and accuracy [58–60]. Bhayani et al. successfully developed a nanoscale complex of dextran-iron oxide nanoparticles which responds to a certain radio frequency [61]. The complex, activated by the certain radio frequency, provides the same changes to the tumour cells as seen from externally heating (43 °C for 60 min) in terms of cell morphology, proliferation patter, and measurement of protein associated with heat shock. Similarly, magnetic nanoparticles can also be used in hyperthermia where the magnetic nanoparticles generate heat after an alternating magnetic field is applied. For example, Sadhukha et al. successfully achieved a significant in vivo inhibition of lung tumour growth by using super-paramagnetic iron oxide nanoparticles (SPIONs) [62].

15.2.2.3 Multifunctional Nanoparticles: Diagnostics and Therapeutics in a Single System

Equipped and stimulated by the rapid development of advanced nanotechnology, it is able to fabricate a nanoscale complex which has multifunctionalities, such as both therapeutic and diagnostic features. A selection of techniques such as nanofabrication, nanoencapsulation, surface grafting and layer-by-layer coating can transform simple nanoparticles to all-purpose multifunctional nanoparticles or nanoplatforms. Multifunctional nanoparticles have been established aiming to improve the particle's stability, biocompatibility, half-life, and add miscellaneous properties (e.g. stimuli responsive, target-specific, disease detecting, or imaging). Particle functionalisation is normally achieved by surface modification [63, 64].

For cancer treatment, attempts have been made to try to create a stable, safe and biocompatible nanoscale system which is capable of (1) accurately targeting tumour cells or tissues; (2) releasing therapeutic agent or performing appropriate treatment in a controlled manner to destroy the tumour cells directly or inhibit their growth; and (3) safely self-degrading or getting itself out of the body through body clearance mechanism. Usually, the system (e.g. non-hybrid multifunctional nanoparticles) consists of a nanoparticle core, shell(s), and surface ligands. non-hybrid multifunctional nanoparticles. Recently, a new class of multifunctional nanoparticles has been created by combining more than one nanomaterials as the system's backbone [63]. They are classified as hybrid multifunctional nanoparticles which possess properties of different backbone materials. Furthermore, they seem to offer possible solutions to current limitation from the non-hybrid systems in terms of the suspension and size stability (once administered into the body), encapsulation effectiveness, controlled release mechanisms, and biocompatibility issues (multi drug resistance and blood compatibility).

Cheng et al. recently a multifunctional system of upconversion nanoparticles (e.g. nanoparticles of lanthanide elements) offering in vivo dual medical imaging: fluorescence and MRI [65]. The system also has the magnetic targeting ability which can increase its accumulation on tumour sites by approximately eight times when compared to the system without the presence of magnetic element. In addition, the system is capable for hyperthermia via near-infrared (NIR) light stimulating which is its therapeutic feature, specific to cancer. Another recent example of multifunctional nanoparticles for the treatment of other diseases was reported by Lee et al. who successfully fabricated targeted gold half-shell nanoparticles for chemo-photothermal therapy of rheumatoid arthritis [66]. Arginine-glycine-aspartic acid (RGD) was conjugated to the nanoparticles for its rheumatoid arthritis-specific targeting ability. The system was loaded with methotrexate, the most effective drug of choice for treating rheumatoid arthritis. The gold nanoparticles further provide hyperthermia ability. When stimulated with NIR radiation, those gold nanoparticles generate heat effectively. The generated heat acts as a trigger for both drug release and direct hyperthermia to cure the diseased sites inside the body. This system further greatly increased the therapeutic efficiency for arthritis using a dramatically reduced dose of methotrexate and its side effects.

15.2.2.4   Nanotechnology in Regenerative Medicine

Regenerative medicine has been increasingly exploited and developed where nanotechnology is utilised in cell therapy, in vivo real-time labelling and imaging, 2D-nanotopography, 3D-nanoscaffold, and growth factor delivery [67, 68]. Nanomaterials have been investigated for their effectiveness, in mechanical, chemical and biological aspects, for making regenerative scaffolds. Furthermore, nanotechnology is able to create opportunities for scientists to develop biomaterials which can mimic various types of extracellular matrices in tissues, generating suitable conditions for triggering cell repair or growth [2, 69].

However, further studies are still required to develop nanomaterials used for regenerative medicine and investigate if they possess the following conditions [2]:

- Non-toxic
- Biocompatible
- Simultaneously facilitate regeneration
- Maintain physical properties (even after being conjugated at the surface)
- Interact with desired target (protein or cell) but not disturbing its normal biological activities

The ultimate goals of regenerative medicine are to induce cell or tissue repairmen without causing other complication from immunological respond or dependence of donors [67].

15.2.2.5 Nanomedicine in Gene Therapy

Gene therapy is defined as a method which utilises appropriate genetic materials, such as fragments of DNA or RNA, to selectively repair faulty genes causing diseases [70]. It is seen as a promising solution as a cure to diseases which are currently incurable or difficult to be cured, such as genetically inherited disorders, certain types of cancer, and viral infections [71]. There are several approaches to cure diseases using different techniques in gene therapy. The most common approach is to replace a non-functional gene on a specific location with a normal gene. Therapy target can be set on the faulty gene too. Homologous recombination can be used to swap a problematic faulty gene with a normal healthy gene. Alternatively, selective reverse mutation on the faulty gene can result in gene repair and the gene is turned into a normal gene [70].

To carry out gene therapy, a messenger called 'gene vector' is required to deliver desired genetic material into the nucleus. Gene therapy starts with the transfection of the target cells by the gene vectors. Genetic materials inside the vectors are then released into the cell. After the genetic materials pass through nuclear membrane into the nucleus, the desired proteins can be synthesised, which will ultimately bring the cell back to its normal condition. There are usually two main types of gene vectors: viral vectors and non-viral vectors. At present, viral gene vectors have a much higher rate of successful delivery compared with non-viral vectors. However, there are issues and concerns using viruses inside the body, despite the fact that they are genetically modified to contain only the desired human genetic materials for the therapy. As a foreign body, it would activate the body's immunological response and even inflammation. Concerns have been raised regarding the fact that there has been no solid evidence to approve: after being administered into the body, the modified viruses would be able to resume their original pathogenic activities and not to cause complication to the patient [70]. As a result, efforts have been continuously made to apply safe and effective non-viral gene vectors. However non-viral vectors have several limitations which are needed to be overcome: stability in biological condition, extracellular obstructions, intracellular obstructions and targeted delivery [72, 73].

Recently nanotechnology has been applied to gene therapy in order to overcome these limitations by creating a stable, safe, and biocompatible non-viral gene vector for effective targeted intracellular gene delivery. Both inorganic and organic (biodegradable) nanoparticles/nanomaterials have been used to create non-viral gene vectors for gene therapy.

Labhasetwar and Panyam described a successful escape of gene-encapsulated PLGA nanoparticles from endosome into the cytoplasm [74]. Yamashita et al. made a photothermally controlled gene delivery system by conjugating double-strand DNA onto gold nanorods [75]. The system has a unique thermal conductivity in response to near-infrared radiation, causing the release of single-strand DNA in a controlled manner. A system of chloroquine-encapsulated polycationic mesoporous silica nanoparticles containing siRNA was shown to have successfully delivered both siRNA and chloroquine [76]. In addition, Chen et al. reported an up-to 50 % gene silencing ability after 48-h post-administration of chitosan-siRNA nanoparticles in

PLGA nanofibers [77]. An approximately 1 week of sustained therapeutic genetic expression was observed by Kwon et al. employing a complex of DNA and cationic lipid-based nanoemulsion [78].

15.3 What Lessons Nanomedicine Can Learn from Nanotechnology?

Nanomedicine incubates and develops from nanotechnology. Broadly speaking, nanomedicine could be regarded as one of the divisions of nanotechnology. Thus, all matters inherited and/or transferred from nanotechnology would certainly affect the progress of nanomedicine.

Lessons, depending on their nature, influence and impact, could be either positive or negative. In the previous sections of this chapter, detailed discussions and sufficient examples are given to demonstrate the positive lessons and results from nanotechnology. It has shown that nanotechnology has broken down several crucial barriers previously existed in medicine, bringing the chance and reality of curing hopeless diseases. It also has improved effectiveness of current medical technologies, such as medical imaging, by enhancing the efficiency of the contrast agents or creating a novel approach of multi-modal medical imaging. Furthermore, the appropriate use of unique physicochemical properties of nanoparticles/nanomaterials enables modulation and control of biological activity at nanoscale, which has been proved to be a promising approach in drug delivery, regenerative medicine, gene therapy and other innovative medical treatments. The following table (Table 15.4)

Table 15.4 Summary of current positive and negative aspects in nanomedicine

Positive aspects	Negative aspects
Smaller devices, hence they are less invasive	Smaller devices require sophisticated technology which may not be economically feasible to everybody
Nanomedical diagnostic devices and methods only require a small amount of sample	Several pre-analysis preparations of the sample are required
Comes in small size and operates at the same level as interaction inside the body	Early developed NPs might accidentally affect various biological barriers, hence results in unexpected toxicities
	Difficult to monitor exposure from outside
Drug delivery technology using nanoparticles protects the drug from being degraded inside the body	The used nanoparticle might be difficult to degrade or come out of the body via normal excretion pathways or mechanisms
NPs-based drugs come in small quantities but with increased efficiency	Damages could be done to healthy tissues or cells if NPs accidentally accumulate at the unexpected areas
Cheaper	More expensive
More accessible to general public	Limited access to affordable population only

summarises the current positive and negative aspects in nanomedicine, leading to a further discussion next on the constructive lessons that nanomedicine can learn from nanotechnology.

15.3.1 Inconsistency Issues in Nanofabrication

As mentioned earlier, there are issues and concerns with respect to the inconsistency of particle sizes and properties, even within the same batch production. The inconsistency could greatly reduce the product quality, and further especially affect the safety of product to be effectively used inside human bodies since it is difficult to confidently control or monitor the effectiveness of the treatment which is not uniform. Serious undesired complications might even occur from this uncontrolled variation.

15.3.2 Industrial Scale-up and Commercialisation Limitations

The techniques and methods evolving nanotechnology could be very specific and expensive. Some of them are also limited to only batch production. Furthermore, some synthetic routes of nanoparticles and other nanomaterials could be too sophisticated so that the product yield might be relatively low and that overall product cost is high. These issues have been preventing several nanotechnology related products from being industrially scaled-up and commercialised as they are not economically feasible with reasonable profitability. It is suggested that researches should be carried out in order to find better and optimised (scaled-up and economic) synthetic routes. Generally speaking, an ideal synthetic route of nanoparticles and other nanomaterials for industrialisation should have the following features:

- Safe
- Easily accessible raw materials
- Including sophisticate steps or equipment as least as possible
- Reasonable capital and operating costs
- Reasonable yield
- Reasonable production timescale
- Useful by-products (if any)

15.3.3 Complex Nature of Biological Phenomena: Additional Complication to the Complex Concept of Nanoscience, Nanotoxicity and Related Matters

Biological phenomena are much more complicated than physical phenomena. Several activities occur simultaneously in a set manner in the biological environment. This is an additional complication to the study of nanoscale phenomena.

Nanotechnology is already a very broad, emerging and extensive field. As discussed earlier, it is somehow difficult to set topics or areas of nanomedicine into discrete categories since everything seems link together.

A critical limitation to the development of nanomedicine at present is the lack of thorough and well-established knowledge regarding interaction between nanomaterials and biological environment inside and outside the body. It has raised public concerns about safety issues regarding the use of nanotechnology in health care, and this is why the study of nanotoxicology will play an more and more important role [2, 79, 69]. The classic examples regarding the nanotoxicity issues include the cases of using silver nanoparticles and carbon nanotubes. Silver nanoparticles have many interesting properties involving the anti-bacterial property which has received extensive interests and investigations. However, it was discovered later that silver nanoparticles has undesired adverse effect in vivo [80]. For carbon nanotubes, it was once at the centre of attraction as a promising drug delivery method and platform. However, it was also discovered later that it could become harmful [81]. Concerns have also been raised with respect to the nanomaterials' fate inside the body and body clearance. In theory, nanomaterials for clinical application has to undergo absorption, distribution, metabolism and excretion (ADME) studies first. However, since it is heterogeneous in content with size distribution, it is difficult to describe ADME properties of nanomaterials [79].

15.3.4 Establishment of Standards and Protocols for Nanomaterial Characterisation

Despite the amazing developments in nanotechnology, there are still limitations of achieving accurate and reliable characterisation of the physical and chemical properties of nano-products. Furthermore, there is no solid standards and protocols for a full/comprehensive and reliable nanomaterial (medical nanomaterial in particular) characterisation mainly due to the diversity, complexity and uncertainty of nanomaterials. There have been suggestions to first develop ex vivo studies of activities of nanoparticles and the body thoroughly from the possible ways of administration, their routes and journey inside the body, and their fate, before performing in vivo studies [11]. However, more issues will rise due to the complexity in fabricating ex vivo system for the study. It is also essential to trace if the nanoparticles are degraded, excreted or accumulated inside the body, as it will affect the toxicity of such particles [79].

15.3.5 Nanomedicine Regulating Bodies: A Demand for Proper Regulations

Nanomedicine has recently come into our visions attributed to the rapid and exciting development of nanotechnology. Researchers are even more optimistic for the future of nanomedicine but it also consequently brings us a new task of determining how to best regulate it so that it is both safe and effective. Similarly occurred during the development of biotechnology, the national governments/bodies worldwide are now struggling with balancing the competing benefits and risks of nanotechnology in the medical and other sectors. It is becoming increasingly clear that reasonable, effective and predictable regulatory structures will be critical to the successful implementation of nanotechnology [82]. When it comes to developing a regulation plan for nanomedicine, the focus needs to be on who will be given the responsibility to oversee regulation and whether to operate under the current regulations or write new regulations [83].

Hence, there have been some discussions on the challenge and suitable role of regulatory bodies such as FDA [84, 85]. It was suggested that whether the FDA should at least look at nanoproducts on a case-by-case basis and should not attempt regulation of nanomedicine by applying existing statutes alone, especially where scientific evidence suggests otherwise. Incorporating nanomedicine into the current regulatory scheme is a poor idea. Hence, regulation of nanomedicine must balance innovation and R&D with the principle of ensuring maximum public health protection and safety. The FDA should also consider implementing several reforms to ensure that it is adequately prepared to regulate nanomedicine.

Chowdhury further recently discussed the regulation of nanomedicine in Europe [86]. Due to the fact that the nanomedicine market in EU is poised at a critical stage wherein clear regulatory guidance is lacking in providing for clarity and legal certainty to manufacturers of nanomedicine, it is imperative to establish suitable regulatory structures for nanomedicine. It was suggested that both the pediatric and the advanced therapies medicinal products regimes offer important regulatory guidance that could be adopted for the regulation of nanomedicines in the EU first.

15.3.6 Ethical Issues

For nanomedicine, an evolved area from nanotechnology receiving increasing attentions from our society, it is crucial to proactively address the ethical, social and regulatory aspects of nanomedicine to minimize its adverse impacts on the environment and public health and to avoid a public backlash [87]. The most significant concerns involve risk assessment, risk management of engineered nanomaterials, and risk communication in clinical trials. Other concerns have been raised regarding privacy violation from generating genetic data of the patient and social justice. Accessibility to health care is also an important issue. Due to generally high-cost of the nanotechnology related products, economic and equity issues have also been pinpointed that only a few people who are financially capable can access this high-end health care but the diseases do not selectively occur in rich population [2, 87]. Educating members of society about the benefits and risks of nanomedicine is thus important to gain and maintain public support.

15.4   Conclusions and Future Outlook

Nanomedicine bridges the gap between nanotechnology and medicine. It is a truly interdisciplinary science which requires cooperation and contribution from engineers, scientists and medical staffs to appropriately, effectively, safely and successfully apply nanotechnology in medicine to move to the next generation of health care. Several discoveries and achievements in nanomedicine have been made, making significant medical advancement and bringing medicine closer to the new era. However, in order to drive nanomedicine further in the right direction with confidence, lessons from nanotechnology must be considered and should be learnt and used wisely to overcome problems and limitations. Complete knowledge and understanding of nanoscale phenomena, such as interaction between nanoparticles and biological environment, nanotoxicity, and nanomaterial physical and chemical properties characterisation are all needed to better refine and catalyse the successful implementation of nanomedicine. It is also very crucial to establish suitable regulating bodies for controlling and monitoring the use of nanotechnology and its products, as well as for providing clarity and legal certainty to manufacturers. The new era of nanomedicine is coming and the potential of nanomedicine seems infinite along with more public awareness and support.

References

1. Logothetidis S (ed) (2012) Nanomedicine and nanobiotechnology. Nanoscience and nanotechnology. Springer, Heidelberg
2. Boisseau P, Loubaton N (2011) Nanomedicine, nanotechnology in medicine. C R Phys 12:620–636
3. Fattal E, Tsapis N (2014) Nanomedicine technology: current achievements and new trends. Clin Transl Imaging 2(1):77–87
4. Tong S, Fine E, Lin Y et al (2014) Nanomedicine: tiny particles and machines give huge gains. Ann Biomed Eng 42(2):243–259
5. Jain K (2008) The handbook of nanomedicine, 1st edn. Humana Press, Totowa
6. Kalangutkar P (2014) The evolution of nanomedicine with the re-evolution of nanotechnology. Int J Eng Sci Invent 3(5):12–16
7. Zarzycki A (2014) Editorial: at source of nanotechnology. Tecno Lógicas 17(32):9–10
8. Vogel V (ed) (2009) Volume 5: Nanomedicine. Nanotechnology. Wiley, Weinheim
9. European Technology Platform on Nanomedicine (2009) Roadmaps in nanomedicine towards 2020 (version 1.0). www.etp nanomedicine.eu/public/.../091022_ETPN_Report_2009.pdf. Accessed on 14 Jan 2014
10. European Technology Platform on Nanomedicine (2013) Roadmaps in nanomedicine towards 2020 (version 1.0). www.etp-nanomedicine.eu/public/.../091022_ETPN_Report_2009.pdf Accessed on 14 Jan 2014
11. Engstrom D, Savu V, Zhu X et al (2011) High throughput nanofabrication of silicon nanowire and carbon nanotube tips on AFM probes by stencil-deposited catalysts. NanoLett 11(4):1568–1574
12. Lieber C (2011) Semiconductor nanowires: a platform for nanoscience and nanotechnology. MRS Bull 36(12):1052–1063
13. Smith GB & Granqvist CGS (2010) Green Nanotechnology: Solutions for Sustainability and Energy in the Built Environment. CRC Press Print ISBN: 978-1-4200-8532-7. eBook ISBN: 978-1-4200-8533-4. http://www.crcnetbase.com/isbn/9781420085334

14. Juliano R (2012) The future of nanomedicine: promises and limitations. Sci Public Policy 39(1):99–104
15. Xu B, Yan X, Zhang J et al (2012) Glass etching to bridge micro- and nanofluidics. Lab Chip 12(2):381–386
16. Saito T, Ohshima S, Xu W et al (2005) Size control of metal nanoparticle catalysts for the gas-phase synthesis of single-walled carbon nanotubes. J Phys Chem B 109(21):10649–10652
17. França R, Zhang F, Veres T et al (2013) Core–shell nanoparticles as prodrugs: possible cyto-toxicological and biomedical impacts of batch-to-batch inconsistencies. J Colloid Interface Sci 389(1):292–297
18. Segerink L, Eijkel J (2014) Nanofluidics in point of care applications. Lab Chip. doi:10.1039/c4lc00298a
19. Stone H, Kim S (2001) Microfluidics: basic issues, applications and challenges. Am Inst Chem Eng J 47(6):1250–1254
20. Wang L, Fan J (2010) Nanofluids research: key issues. Nanoscale Res Lett 5:1241–1252
21. Liu L, Yoo S, Lee S et al (2011) Wet-chemical synthesis of palladium nanosprings. Nano Lett 11:3979–3982
22. Choi C, Kim C (2006) Fabrication of a dense array of tall nanostructures over a large sample area with sidewall profile and tip sharpness control. Nanotechnology 17(21):5326–5333
23. Huefner S (2006) Nanobiosensors. http://www.chem.usu.edu~tapaskar/Sara.ppt. Accessed on 1 July 2013
24. Topal C (2011) Nanobiosensor. http://bionanotech.uniss.it/wpcontent/uploads/2011/09/bio-sensori.ppt. Accessed 1 July 2013
25. Saha K, Agast S, Kim C et al (2012) Gold nanoparticles in chemical and biological sensing. Chem Rev 112:2739–2779
26. Cao X, Ye Y, Liu S (2011) Gold nanoparticle-based signal amplification for biosensing. Anal Chem 417(1):1–16
27. Pumera M, Sanchez S, Ichinose I et al (2007) Electrochemical nanobiosensors. Sens Actuators B 123:1195–1205
28. Steffens C, Leite F, Bueno C et al (2012) Atomic force microscopy as a tool applied to nano-biosensors. Sensors 12:8278–8300
29. Tang B, Cao L, Xu K et al (2008) A new nanobiosensor for glucose with high sensitivity and selectivity in serum based on fluorescence resonance energy transfer (FRET) between CdTe quantum dots and Au nanoparticles. Chem Eur J 24:3637–3644
30. Lad A, Agrawal Y (2012) Optical nanobiosensor: a new analytical tool for monitoring carbo-platin–DNA interaction in vitro. Talanta 97:218–221
31. Elahi M, Bathaie S, Mousavi M et al (2012) A new DNA-nanobiosensor based on g-quadruplex immobilized on carbon nanotubes modified glassy carbon electrode. Electrochim Acta 82:143–151
32. Parolo C, Merkoci A, Mousavi M et al (2013) Paper-based nanobiosensors for diagnostics. Chem Soc Rev 42:450–457
33. Hoshino Y, Takashi K, Yoshio O, & Kenneth JS (2008) Peptide imprinted polymer nanopar-ticles: a plastic antibody. J Amer Chem Soc 130(46):15242–15243
34. Foudeh A, Didar T, Veres T et al (2012) Microfluidic designs and techniques using lab-on-a-chip devices for pathogen detection for point-of-care diagnostics. Lab Chip 12(18):3249–3266
35. Medina-Sanchez M, Miserere S, Merkoci A (2012) Nanomaterials and lab-on-a-chip technolo-gies. Lab Chip 12:1932–1943
36. Schumacher, Soeren, Jörg Nestler, Thomas Otto, Michael Wegener, Eva Ehrentreich-Förster, Dirk Michel, Kai Wunderlich et al (2012) 'Highly-integrated lab-on-chip system for point-of-care multiparameter analysis." Lab Chip 12(3):464–473
37. Govindarajan A, Ramachandran S, Vigil G et al (2011) A low cost point-of-care viscous sam-ple preparation device for molecular diagnosis in the developing world: an example of micro-fluidic origami. Lab Chip 12(1):174–181
38. Carmode D, Skjaaa T, Fayad Z et al (2009) Nanotechnology in medical imaging: probe design and applications. Arterioscler Thromb Vasc Biol 29:992–1000

39. Lee N, Choi S, Hyeon T (2013) Nano-sized CT contrast agents. Adv Mater 25(19):2641–2660
40. Chien C, Chen H, Lai S et al (2012) Gold nanoparticles as high-resolution X-ray imaging contrast agents for the analysis of tumor-related micro-vasculature. J Nanobiotechnol 10:10
41. Bae H, Ahmad T, Rhee I, Chang Y, Jin SU, Hong S (2012) Carbon-coated iron oxide nanoparticles as contrast agents in magnetic resonance imaging. Nanoscale Res Lett 7:44
42. Gupta A, Arora A, Menakshi A et al (2012) Nanotechnology and its applications in drug delivery: a review. WebmedCentral Med Educ 3(1):MC002867
43. Liu H, Liu T, Wang H et al (2013) Impact of PEGylation on the biological effects and light heat conversion efficiency of gold nanoshells on silica nanorattles. Biomaterials 34(28):6967–6975
44. Wang M, Thanou M (2010) Targeting nanoparticles to cancer. Pharmacol Res 62:90–99
45. Hatakeyama H, Akita H, Ishida E et al (2007) Tumor targeting of doxorubicin by anti-MT1-MMP antibody-modified PEG liposomes. Pharm Nanotechnol 342(1–2):194–200
46. Hammond P (2010) Thin films: particles release. Nat Mater 9:292–293
47. Dubey V, Mishra D, Nahar M et al (2010) Enhanced transdermal delivery of an anti-HIV agent via ethanolic liposomes. Nanomed Nanotechnol Biol Med 6(4):590–596
48. Mulik R, Mönkkönen J, Juvonen R et al (2010) Transferrin mediated solid lipid nanoparticles containing curcumin: enhanced in vitro anticancer activity by induction of apoptosis. Int J Pharm 398(1–2):190–203
49. Goldberg D, Vijayalakshmi N, Swaan P et al (2011) G3.5 PAMAM dendrimers enhance transepithelial transport of SN38 while minimizing gastrointestinal toxicity. J Control Release 150(3):318–325
50. Guo J, Gao X, Su L et al (2011) Aptamer-functionalized PEG–PLGA nanoparticles for enhanced anti-glioma drug delivery. Biomaterials 32(1):8010–8020
51. Makadia H, Siegel S (2011) Poly lactic-co-glycolic acid (PLGA) as biodegradable controlled drug delivery carrier. Polymers (Basel) 3(3):1377–1397
52. Sahni J, Baboota S, Ali J (2011) Promising role of nanopharmaceuticals in drug delivery. Pharma Times 43(10):16–18
53. Savla R, Taratula O, Garbuzenko O et al (2011) Tumor targeted quantum dot-mucin 1 aptamer-doxorubicin conjugate for imaging and treatment of cancer. J Control Release 153(1):16–22
54. Taghdisi S, Lavaee P, Ramezani M et al (2011) Reversible targeting and controlled release delivery of daunorubicin to cancer cells by aptamer-wrapped carbon nanotubes. Eur J Pharm Biopharm 77(2):200–206
55. Saad Z, Jahan R, Bagal U (2012) Nanopharmaceuticals: a new perspective of drug delivery system. Asian J Biomed Pharm Sci 2(14):11–20
56. McDowell G, Slevin M, Krupinski J (2011) Nanotechnology for the treatment of coronary in stent restenosis: a clinical perspective. Vasc Cell 3:8
57. Tan A, Alavijeh M, Seifalian A (2012) Next generation stent coatings: convergence of biotechnology and nanotechnology. Trends Biotechnol 30(8):406–409
58. American Cancer Society (2011) Hyperthermia. http://www.cancer.org/treatment/treatmentsandsideeffects/treatmenttypes/hyperthermia. Accessed 29 June 2013
59. Baronzio G, Parmar G, Ballerini M et al (2014) A brief overview of hyperthermia in cancer treatment. J Integr Oncol 3(1):1–10
60. Di Corato R, Espinosa A, Lartigue L et al (2014) Magnetic hyperthermia efficiency in the cellular environment for different nanoparticle designs. Biomaterials 35(24):6400–6411
61. Bhayani K, Rajawade J, Paknikar K (2013) Radio frequency induced hyperthermia mediated by dextran stabilized LSMO nanoparticles: in vitro evaluation of heat shock protein response. Nanotechnology 24(1):015102
62. Sadhukha T, Wiedmann T, Panyam J (2013) Inhalable magnetic nanoparticles for targeted hyperthermia in lung cancer therapy. Biomaterials 34(21):5163–5171
63. Morales C, Valencia P, Thakkar A et al (2012) Recent developments in multifunctional hybrid nanoparticles: opportunities and challenges in cancer therapy. Front Biosci 4:529–545
64. Lee D, Koo H, Sun I et al (2012) Multifunctional nanoparticles for multimodal imaging and theragnosis. Chem Soc Rev 41:2656–2672

65. Cheng L, Yang K, Li Y et al (2012) Multifunctional nanoparticles for upconversion lumines-cence/MR multimodal imaging and magnetically targeted photothermal therapy. Biomaterials 33(7):2215–2222

66. Lee S, Kim H, Ha Y et al (2013) Targeted chemo-photothermal treatments of rheumatoid arthritis using gold half-shell multifunctional nanoparticles. ACS Nano 7(1):50–57

67. Bean A, Tuan R (2013) Stem cells and nanotechnology in tissue engineering and regenerative medicine. In: Ranalingam M, Jabbari E, Ramakrishna S et al (eds) Micro and nanotechnolo-gies in engineering stem cells and tissues, 1st edn. Wiley, Hoboken

68. Kingham E, Oreffo R (2013) Embryonic and induced pluripotent stem cells: understanding, creating, and exploiting the nano-niche for regenerative medicine. ACS Nano 7(3):1867–1881

69. Thierry B, Textor M (2012) Nanomedicine in focus: opportunities and challenges ahead. Biointerphases 7(19)

70. Human Genome Project Information (2011) Gene therapy. http://ornl.gov/sci/techresources/ Human_Genome/medicine/genetherapy.shtml. Accessed 2 July 2013

71. Genetics Home Reference (2013) What is gene therapy? http://ghr.nlm.nih.gov/handbook/ therapy/genetherapy. Accessed 2 July 2013

72. Labhasetwar V (2005) Nanotechnology for drug and gene therapy: the importance of under-standing molecular mechanisms of delivery. Curr Opin Biotechnol 16(6):674–680

73. Gascón A, Pozo-Rodríguez A, Solinís M (2013) Non-viral delivery systems in gene therapy, gene therapy – tools and potential applications, Dr. Francisco Martin (Ed.), ISBN: 978-953-51-1014-9, InTech, DOI: 10.5772/52704. Available from: http://www.intechopen.com/books/ gene-therapy-tools-and-potential-applications/non-viral-delivery-systems-in-gene-therapy

74. Labhasetwar V, Panyam J (2003) Biodegradable nanoparticles for drug and gene delivery to cells and tissue. Adv Drug Deliv Rev 55(3):329–347

75. Yamashita S, Fukushima H, Akiyama Y et al (2011) Controlled-release system of single-stranded DNA triggered by the photothermal effect of gold nanorods and its in vivo applica-tion. Bioorg Med Chem 19:2130–2135

76. Bhattara S, Muthuswamy E, Wani A et al (2010) Enhanced gene and siRNA delivery by polycation-modified mesoporous silica nanoparticles loaded with chloroquine. Pharm Res 27(12):2556–2568

77. Chen M, Gao S, Dong M et al (2012) Chitosan/siRNA nanoparticles encapsulated in PLGA nanofibers for siRNA delivery. ACS Nano 6(6):4835–4844

78. Kwon S, Nam H, Nam T et al (2008) In vivo time-dependent gene expression of cationic lipid-based emulsion as a stable and biocompatible non-viral gene carrier. J Control Release 128(1):89–97

79. Nystrom A, Fadeel B (2012) Safety assessment of nanomaterials: implications for nanomedi-cine. J Control Release 161:403–408

80. Han DW, Woo YI, Lee MH, Lee JH, Lee J, Park JC (2012) In-vivo and in-vitro biocompatibil-ity evaluations of silver nanoparticles with antimicrobial activity. J Nanosci Nanotechnol 12(7):5205–5209

81. Zhao Y, Xing G, Chai Z (2008) Nanotoxicology: are carbon nanotubes safe? Nat Nanotechnol 3(4):191–192

82. Marchant GE, Sylvester DJ, Abbott KW, Danforth TL (2010) International harmonization of regulation of nanomedicine. Stud Ethics Law Technol 3(3):1941–6008

83. Shanna H (2009) The Regulation of Nanomedicine: Will the Exisiting Regulatory Scheme of the FDA Suffice?, XVI Rich. J.L. & Tech. 4 http://law.richmond.edu/ jolt/v16i2/article4.pdf

84. Miller J (2003) Beyond biotechnology: FDA regulation of nanomedicine. Columbia Sci Technol Law Rev 4:E5

85. Bawa R (2011) Regulating nanomedicine – can the FDA handle it? Curr Drug Deliv 8(3):227–234

86. Chowdhury N (2010) Regulation of nanomedicines in the EU: distilling lessons from the pedi-atric and the advanced therapy medicinal products approaches. Nanomedicine 5(1):135–142

87. Resnik DB, Tinkle SS (2007) Ethics in nanomedicine. Nanomedicine 2(3):345–350

Chapter 16
Intersection of Nanotechnology and Healthcare

Swasti Gurung, Dan Fei, and Yi Ge

16.1 Introduction – Brief Background of Nanotechnology

Nanotechnology is the engineering of functional systems and manufacturing of materials at atomic and molecular scale. Modern word *nanotechnology* is derived from Greek term 'nano' meaning dwarf. Dr. Richard Feynman, the Nobel Prize winner in Physics in 1959, is widely recognised as the "father" of the subject, but the term *nanotechnology* was formally introduced by Professor N. Taniguchi in the year 1974 who defined nanotechnology as "the processing of separation, consolidation and deformation of materials by one atom or one molecule" [1]. Nano in the International Systems of Units (SI) is known as one billionth of a meter (10^{-9} m). The size of a structure to be classified as 'nano' usually needs to be roughly between 1 and 100 nm (nano-meter) in size at any one dimension. Regardless of the size restriction, nanotechnology implies to any structures that are developed by top-down or bottom-up engineering of individual components even if it is several hundred nanometers in size [2]. The top-down fabrication is normally applied for achieving nanometric precision and accuracy in an artefact by material removal or

S. Gurung • Y. Ge (✉)
Centre for Biomedical Engineering, Cranfield University,
Cranfield, Bedfordshire, MK43 0AL, UK
e-mail: y.ge@cranfield.ac.uk

D. Fei
Leicester School of Pharmacy, De Montfort University,
The Gateway, Leicester, LE1 9BH, UK

© Springer Science+Business Media New York 2014
Y. Ge et al. (eds.), *Nanomedicine*, Nanostructure Science and Technology,
DOI 10.1007/978-1-4614-2140-5_16

deposition (e.g. nanolithography and energy beams). In comparison, the bottom-up fabrication is a technique used for assembling components of nanometric or sub-nanometric size in creation of an artefact with functions such as molecular assembly or manipulation. According to the National Nanotechnology Initiatives (NNI), defining features of nanotechnology are as follows [3]:

- Any research and technology involving development at 1–100 nm range.
- Creation and usage of structures that have novel properties because of their size.
- Ability to control and manipulate at atomic scale.

Nanometre sized objects possesses remarkable and noble properties such as self-assembly and self-ordering under control of forces which macro objects are not capable of doing [4]. The unique features of nano objects make nanotechnology feasible for altering and improving properties and performances of conventional objects and products.

16.2   Healthcare

Healthcare is the diagnosis, treatment, and prevention of disease, illness, injury, and other physical and mental impairments in humans. It is one of those sectors in any place of any country in the world that is always looking for ways to improve and out perform their existing records. Healthcare is also among the highly invested sectors publically or privately in most of the countries since the health of a citizen is directly related to the country's level of development and progress. According to the official information supplied by the National Health Service (NHS) which is one of the largest publically funded services in the world, around £108.9 billion was budgeted for the financial year of 2012/13 in the UK [5]. This shows the calibre of healthcare investment required in order for it to function smoothly. It also shows the potential and opportunity it holds to cut down the running cost yet maintaining the quality of service or even improving it.

Nanotechnology comes into association with healthcare in various known or unknown forms. For example, one of the most well-known or prominent services where nanotechnology could contribute and aid is the imaging service/technology, such as X-rays, MRI (Magnetic Resonance Imaging) scan, and CT (Computerised Tomography) scan. According to the statistical data released from the Department of Health in the UK, the total number of imaging examinations or tests was 40.1 million during the period of 1st April 2011 to 31st March 2012 (DH, [6]). That is a total of 3.3 % (1.3 million tests) increase in the amount of examinations performed in comparison to previous year 2010–2011. Among those, X-rays (radiographs) totalled to 22.5 million, ultrasound of 9.0 million, CT scan of 4.4 million, MRI scan of 2.3 million, fluoroscopy of 1.3 million and 0.6 million were radio-isotopes. This is just an example of a fraction of the entire healthcare service provided but the amount is ample enough to demonstrate the scale of operation.

16.3   Advances of Nanotechnology in Healthcare

From engineering to designing, environmental science to textile manufacturing, many fields are able to change and benefit from a novel technology, such as nanotechnology, that is garnering pace and success at an incredible speed. Healthcare is no stranger to such revolutionary technology either and the conjugation of nanotechnology in relation to medicine is favourably named 'nanomedicine'. Nanomedicine has brought many commendable advancements in healthcare and what seemed like scientific fiction a few years back is now almost within grasp of reality. Among various fields of medicine, the pharmaceutical industry is the most adaptive, evolutionary and fast paced one getting into holds with precognitive and innovative idea such as nanotechnology, and has made a significant breakthrough as well. Nanomedicine at present is a profound part of pharmaceutical industry and is making a major breakthrough in medicine.

16.3.1   Applications of Nanotechnology in Imaging

Molecular imaging is an emerging technology which intends to improve the accuracy of disease diagnosis, and the association of nanotechnology has aided molecular imaging to further have the capacity of disease pathology characterisation. This has been achieved by using a number of nanoparticles (NPs), such as liposomes, dendrimers, gold, iron oxide and perfluorocarbon NPs, which are very sensitively and selectively responsive to specific disease [7, 8]. In addition, in combination with other advantages such as small size, high surface area to volume ratio, long circulating hour, high affinity for target tissue, easy production, less toxicity and immunogenicity, it makes NPs highly attractive to be used as the contrast agents [9, 10]. For example, tailored NPs (e.g. dendrimers) could be used as advanced contrasting agents for MRI scan by shortening the spin–lattice relaxation time T1 and spin-spin relaxation time T2, resulting in sharper and brighter images. Herein, T1 and T2 are two different types of MRI scans that help differentiates types of abnormalities in accordance to density. T1 is effective in imaging solid organ pathology (e.g. liver and spleen), whereas T2 is effective in imaging soft tissues. Furthermore, the NPs (e.g. iron oxide NPs) that have superparamagnetic properties are able to change the spin-spin relaxation time of neighbouring water molecules. Thus, they could be applied to monitor the expression of genes, detect tumours, artherosclerotic plaques, and tissue inflammation, etc. The NPs can also be targeted actively or passively in favour according to the subject of interest to differentiate normal and diseased tissues.

Furthermore, NPs could have multiple binding sites which increases affinity for target tissues immensely [10]. Intracellular imaging is possible with NPs like quantum dots which have high-fluorescence intensity making it easier for tracking of cells throughout the body. They are more desirable then conventional fluorescent since they are more stable, allowing images to be sharper and crisp over long period of time. In addition, they have the ability to detect multiple signals at the same time [11]. They emit bright lights which mean small amount of quantum dots can be sufficient to produce desired signal which makes them promising candidate for detection and diagnosis of various diseases. Due to their small size it is easier for them to enter and interact with biomolecules within the cells.

Atomic force microscopy (AFM) is another powerful tool for imaging which incorporates nanotechnology providing results with high-resolution and three-dimensional images [12]. An AFM functions with a microscale cantilever which has a sharp tip, used as a probe to scan specimen surfaces under focus. Deflection of the probe is detected and recorded as a result which happens due to the presence of attraction and repulsion force between the close contact of probe and surface of the specimen which is normally within 1–10 nm in distance. Deflection of the probe is quantified by the beam bounce method where lasers beam on top of the cantilever into range of photodiodes providing a three-dimensional profile of the specimen on scan. The development progression in AFM has allowed tumour detection, detection of erythrocytes influenced by diabetes and studying the structure of C-reactive protein that are a risk for coronary artery disease and peripheral arterial diseases [12]. Also the application of nanotechnology in molecular imaging is giving ways for further development of specialised medicine like personalised medicine and is sure to bring about a huge revolution and transform the way of disease diagnosis, treatment and prevention.

16.3.2 *Applications of Nanotechnology in Drug Delivery*

When talking about nanotechnology in healthcare, drug delivery system is one of the most conspicuous topics that attract a huge amount of interests from both civil society and industry. Pharmaceutical company's quest to develop targeted drug delivery systems with existing drugs whilst incorporating nanotechnology for effective medical treatment is evolving on daily basis. Effective drug-targeting system based on therapeutic efficacy, appropriate concentration and longer circulation time, could be achieved by utilising nanotechnology [2, 7, 13, 14]. For example, nanoparticles (NPs) could be employed to deliver drugs to specific types of cells (e.g. cancer cells) whilst overcoming barriers such as heat, light and various physiochemical environments. These NPs are engineered in specific way such that it is able to adhere to targeted diseased cells and delivers direct treatment to those cells alone. They could further help to reduce damage to healthy cells significantly, decrease side effects and even allow earlier detection of diseases.

In terms of the improved drug delivery system via a better and innovative formulation, some drug nanocrystals have already been commercially developed [15]. For instance, Elan Nanosystems developed a process called nanonisation to solve the problem of poor water solubility of a drug. This was achieved by reducing the drug crystals until they became particles of 400 nm in diameter or less. A thin layer of polymeric surface modifier was used for absorptions onto the crystal surfaces to prevent aggregation and for stabilisation of the particles produced. The result was a suspension that looked and functioned like a solution which can be used in various forms of dosage such as pills, sprays or creams.

Researchers in Kyoto University developed a smart drug that got activated in specific circumstances [16]. In this case, the novel drug molecule released antibiotic only in presence of an infection. A molecule of gentamicin was bound to a hydrogel with a peptide linker which is cleavable to a proteinase enzyme produced by *Pseudomonas aeruginosa*. *Pseudomonas aeruginosa* is a common bacterium that can cause disease in animals and humans. This smart drug was tested on rats, which showed that in presence of a bacterium, the enzymes produced by the microbes cleaved the linker releasing gentamicin which then killed the bacteria. But in the case where there was no presence of bacteria or the enzyme produced, the drug remained unaffected.

Freitas Jr Robert A. designed a spherical nanorobot the size of a bacterium and made up of 18 billion atoms which were arranged precisely in a crystalline structure to form a miniature pressure tank (Freitas Jr, [17]). The design was of an artificial red blood cell called respirocyte. The miniature tank would hold as much as nine billion oxygen and carbon dioxide molecules. When the artificial blood was to be injected into an individual's bloodstream, the sensors on the surface of the robots would detect the level of oxygen and carbon dioxide in the blood and signal the time to load oxygen and unload carbon dioxide and vice-versa. These nanobots could store and transport gas 200 times more than red blood cells and also consists of glucose engine which releases glucose when there was a deficiency in the body. Even though this is a conceptual idea, there have been uses of artificially engineered microbes already to produce human hormones. Such example is incorporation of human DNA in the genome of bacteria which then starts producing human hormones used for curing endocrine diseases.

16.3.3 Applications of Nanotechnology in Gene Delivery

Nanotechnology has been applied in gene delivery with help of NPs such as liposomes and dendrimers [14, 18, 19]. In able to be successful in this venture, understanding gene therapy is important. Gene therapy is a technique which involves altering, removing or inserting gene at particular loci in order to treat various genetic disorders. In order to do so, a factor able to transfer the gene to a desired location is required which is known as a vector. A vector can be viral or non-viral of origin and mostly includes retroviruses, adenoviruses, lentiviruses and adeno-associated viruses which are very useful in utilising the natural mechanism of an infection [7].

Gene delivery replaces defective gene with a normal one or delivers genes into the disease cells to cure and treat diseases. It was applied as a method to treat hereditary diseases earlier but now have been proven very helpful in treating diseases like cancer. Even then, there are certain limitation points that the technique faces which have been able to overcome by help of nanotechnology, with introduction of non-viral vectors like liposomes and dendrimers which are less immunogenic than the conventional viral vectors. The properties that NPs behold which make them better vectors than currently used vectors are as follows [20, 19]:

- They are cationic in nature and encapsulates negatively charged DNA by electrostatic interactions;
- Safe and simple in use;
- Easily reproducible;
- Even with decreased efficiency in transfection compared to viral vectors, adjustments are easy to make which overcomes the shortcoming.

Dendrimers are known for being efficient in gene delivery and have the ability to protect DNA from the action of DNAse enzyme. The transfection efficiency can be increased by performing heat treatment with solvents like water and butanol which enhances flexibility, allowing dendrimers to become compact when compounded with DNA. The dendrimer that is most commonly used is Polyanidoamine (PAMAM) because it has the highest transfection efficiency [14, 19].

Another such use of NPs, as gene delivery carriers, are liposomes which have certain advantages [19]. Their size can be easily controlled and modified to add a targeting agent but they do have a downfall of having low efficiency in encapsulating DNA. However, the issue of low efficiency can be solved or avoided by using cationic liposomes because they consist of lipid bilayers which are positively charged and combine spontaneously with the negatively charged DNA. The liposomes are mixed with cholesterol and further modified with functional ligands to increase transfection efficiency.

16.4   Impacts of Nanotechnology on Healthcare

The Impact of nanotechnology is extending, from medical, environmental, biology, computing, material science to even communications and military applications. Even though nanotechnology is showing promising development and positive results are being demonstrated steadily, the fact that it is still an emerging field which cannot be blindsided easily. The sole fact that it is an emerging field has roused numerous heated debates about the extent till where the technology benefits and the risks for human health it may bear. Based on the impact of nanotechnology on human health strictly, the subject could be divided into two categories: (1) Potential of nanotechnology holds for innovative medical applications in curing diseases; (2) Potential health hazard it may pose with exposure to nanomaterials. The sustainability of nanotechnology would thus depend on social acceptance, minimised risks and maximised benefits.

The biggest concern related with nanotechnology in regards to its applications in healthcare would be the unknown outcome when exposing to nanomaterials. Due to the scarcity of systematic studies and established regulations, nanotechnology is not easily accepted by many fields even with its promising and positive results.

Healthcare is in no exception since it holds responsibility of millions and billions of people and their health with any small decision they make that has anything to do with treatment, diagnosis and cure. Showing promising results is still not convincing enough for medical society to accept nanotechnology without hesitance because of the technology lacking long documented track record like those of conventional and traditional technologies. Nanotoxicity is being pursued with rigorous pace and through experimentation but it still is not being able to reach the standard required by many healthcare organisations. Especially with the case like asbestosis where the symptoms usually appeared 30-40 years after the exposure period when the damage was catastrophic and unmanageable, the requirement for a more comprehensive investigation of nanotoxicity in relation to nanomaterial exposure is urged and becomes imperative in healthcare.

16.4.1   Nanotoxicity

Materials possess very different properties in comparison to their initial bulk form, such as surface area, surface properties and chemical properties. These different properties have intrigued many scientific innovation and experiments which have made many ground-breaking progress over the years. Nanomaterials have some unique properties in comparison to their larger counterparts due to the quantum size effects and large surface area to volume ratio. Hence, manipulation of substance at nanoscale will have variety of effects in manufacturing, engineering, environmental technology, information technology, health, pharmaceuticals and many other industries. In other words, the resulting nano-sized materials may offer a safe solution or pose a threat to the environment and to human beings [21]. Since there still are no sufficient data available for identifying, monitoring, and controlling the toxicity of nanomaterials, this concern gets brought upon time and again by environmental activist and regulatory bodies, etc. [22].

With respect to our body system in particular, nanomaterials which are very fine could easily be inhaled. They can re-disperse within body to different organs after initial stage of introduction inside the body. Some of the routes which nanomaterials adapted to enter the body are as follows [23, 24]:

- Respiratory system
- Ingestion
- Dermal exposure
- Medical implants (e.g. orthopaedic)

Since there is no cut-off point below about which particles are suddenly classed harmful, in relation to NPs, two factors in mixture may determine the potential harm caused (esp. in lung injury).

1. Large surface area and reactivity of the surface
2. Smaller particles which are more likely to be harmful

Once NPs are inside the human body, they could mix with blood during gas exchange and get transported to different organs of the body during circulation. They could get deposited even in nervous system due to their ability to overcome the blood brain barrier. Collectively, some in vitro studies have already identified that the oxidative stress related changes of gene expression and cell signalling pathways may underlay the toxic effects of NPs [24, 25]. Similar effect in role of transition metals and certain organic compounds on combustion generated NPs were also found [25, 26]. Recently, according to Health and Safety Executive [27] nanomaterials are classified hazardous under following criteria:

- Thinner than 3 μm
- Longer than 10–20 μm
- Biopersistent
- Do not dissolve/break into shorter fibers

16.4.2 Nanopharmaceuticals and Food and Drug Administration (FDA)

Lots of pharmaceutical companies are in trouble with patent expirations on numerous 'blockbuster' drugs, resulting in a loss of multi-billion dollars [28]. There has been an argument over big pharma companies being more focused on shareholder profits than innovative therapies. In today's global economy, big pharmaceutical companies face huge pressure to deliver high-quality products while maintaining profitability. Because of this rising issue, nanotechnology has been applied by numerous pharmaceutical companies to revisit their shelved drugs that were difficult to formulate due to their solubility profiles. The existing nanopharmaceuticals in market that have been approved by the FDA are in absence of any special testing in accordance to the pre-existing laws [28, 29]. However, the approval of new nano-drugs and 'nano-reformulations' has challenged FDA's regulatory framework, which as forced FDA to evaluate submitted products for market approval on the category based-system. A drug, biologic or device has been assigned to Centre for Drug Evaluation and Research (CDER), the Centre for Biologics Evaluation and Research (CBER) or the Centre for Devices and Radiological Health (CDRH) for evaluation respectively. Certain therapies which comprises of two or more components (drug, biologic or a device) that are physically, chemically or otherwise combined or mixed to produce a single entity is 'combination entity'. However, this arrangement has resulted in inconsistency when approved by the FDA in basis of category-based approval which had been deemed "arbitrary and capricious". With issues such as this, nanopharmaceutical is more likely to complicate the combinational products with potential to further blur the lines in distinguishing these categories [29]. In addition, nanopharmaceuticals may also present safety issues for FDA knowing the unpredictable nature of interactions between nanoparticles and biological systems since the surface charge and shape associated with a NP is known to influence its toxicity. Another particular safety issue to be raised by nanopharmaceuticals is the potential for bioaccumulation of NPs with prolonged use [28]. For example, buckminsterfullerene has shown to impair DNA repair mechanism with additional report of certain NPs shown to cause brain damage in fish and lung toxicity in mice.

16.5 Regulatory Challenges to Nanotechnology

Application of existing regulatory frameworks and space for tailoring rules implementing new technologies and products development is questionable when it comes to nanotechnology. This puts pressure on regulators capacity to keep in pace with developments such as nanomedicine and other new applications. Difficulties arise in balancing technological benefits to risks for expertise in regulatory bodies. New innovations such as nanotechnology require practical regulators, who are able to facilitate responsible development in order to gain trust of stakeholders for such areas to prosper.

As a fact, the knowledge gaps in product formulation and concentration of NPs have raised questions about the applicability of European regulations on the Registration, Evaluation, Authorisation and Restriction of Chemicals (REACH) [30]. Another debatable issue involving nanotechnology is to do with uncertainty and ambiguous risk that rise dilemma in regulators on either to wait until there is sufficient knowledge available or to act promptly. Another knowledge gap in nanotechnology is to do with the toxicological aspect of nanotechnology and the potential risk it pose on health and environment, which challenges safety regulations to its limit. As recorded, toxicological studies conducted with NPs have indicated that free NPs could penetrate through the blood–brain barriers or remaining lodged in capillaries [25]. There has been prediction for possible impact of these particles on immune system and consumption by macrophages. Uncertain biocompatibility imposed by NPs in relation to it being used in medical products and materials is another challenge on its own. It has been evaluated that the toxicological risks of NPs depend on material properties, exposure route, dose and frequency of doses which accounts for risks with usage of NPs among other things such as distribution of particles in the body [24, 25]. Even worse, further toxicological risks of NPs are yet to be known. Previous drug disaster in late 1950s with thalidomide, of which effects are still being projected till today, regulation bodies have thereafter ensured to higher the level of public health protection in terms of safety, efficacy and quality.

16.6 Rules Governing Nanomedicine

There is still no specific rule regarding nanomedicine by the European Union to date. Having said that, nanopharmaceuticals have been classed as advanced therapy medicinal products and can only be approved by centralised procedure. In cases where nanomedicinal products are in combination with medical devices and/or regular medical products, certain aspects of regulatory regimes for both medical devices and pharmaceuticals apply, regardless of the manner in which the other features have been combined in the product [30]. The application of regulatory regime depends on the category in which the product falls in regards to the definition of medicinal products, advanced therapies medicinal products or medical devices. The primary mode of action depends on the criteria in which the products falls majorly in application to the regulatory regime. The market authorisation depends on positive outcome of risk-benefit balance. Applicant must be able to demonstrate sufficient product safety, quality and efficacy in comparison to large set of objective scientific data. It is mandatory for scientific evaluation of applications to be based on highest level of expertise and standards.

16.7   Marketing Prospects of Nanotechnology

Nanotechnology is bound to have a substantial impact on the world's economy and market volumes, which are a good indicator for such economic significance. Despite all the controversies and hesitance in its acceptance, if successful the technology will contribute substantially. There were plenty of market forecast originated for nanotechnology during the early 2000s with timeline going up to the year of 2015. Among all, the best compiled forecast has to be the one published by the National Science foundation (NSF) of US in 2001. The NSF forecasted that the estimated world market of nanotechnology would worth 1 trillion US dollars by 2015 [31]. In addition, by combination of other technologies, nano-enabled products and markets are expected to be of the largest share in the world. Nanotechnology has already been attracting significant amount of investments from government and various business communities around many parts of the world. In 2007, it was estimated that the total global investment nanotechnology held was around five billion Euros of which two billion Euros were from private sectors [32].

In addition to the booming nanotechnology assisted market, there was also a remarkable increase in published patents of nanotechnology, which ranged from 531 total patents in 1995-1976 total published patents by 2001 (Royal Society [33]).

16.8   Public's Concern & Prospects on Nanotechnology

It is evident that nanotechnology have bought about remarkable differences in ways of diagnosis, patient care and other medical and non-medical implications, yet it has not been able to establish itself in full positive light within the general public's eye due to the lack of communication in interpretation of it. There are very few studies that have been carried out about the media coverage of nanotechnology [34]. Mass media plays a significant role in shaping public attitude towards nanotechnology or any other field of discovery and development since they are the major source of information to the general public. According to a survey conducted in the US and the UK, some common conclusions were drawn [34]:

- Media interest in nanotechnology has grown immensely since 1999 and in 2003 it began spreading from opinion-leading elite press to the general press hence addressing wider population;
- Media coverage of nanotechnology throughout the period of analysis (1984–2004) has been overwhelmingly positive although there are articles about risks nanotechnology bear;
- Majority of media had presented nanotechnology in terms of progress and economic prospects.

Even when media has been positive towards nanotechnology, the public can be more sensitive to possible impacts of new technologies. A good example of it is the toxicity of NPs. The word 'nano' has been embedded in the national consciousness and is an area of public debate and often concern. From scientific fictional tales of self-replicating 'nanobots' engulfing the word to legit concerns on effect of NPs used in everyday products such as sun creams, it is inevitable for nanotechnology to be out of public view. Factors such as emotive, ethical and political implications also come in to play a major role. One of the best known example of such an issue is the stem cell research. It has managed to gain the highest scientific profiles in both medical community and the general public in the past decade.

Due to the novelty of the technology, a full acceptance is yet to be achieved and acknowledged. Main concern lies in public's view on the development and effective way to convey the novel method. It always is our human nature to have curiosity around new subject and have the expectation to know more about it. There are a few of ways where the message can be conveyed efficiently as follows [34]:

- Learning from previous cases, avoid the same mistakes;
- Aim to elucidate the public's knowledge and attitude towards the technology;
- Public workshops;
- Focus groups;
- Sources that give further information about genuinely considered beliefs of public towards nanotechnology instead of currently uninformed opinions;
- Assessments of nanotechnology as positive and major benefits, especially from health application of nanotechnology;

On the other hand, from the academic point of view, there are also a few of things which could help to dilute the issue and enhance a better future for nanotechnology, including nanomedicine in healthcare [34]:

- Possible ways to deal with inherent uncertainty concerning the potential impacts and future developments;
- Urge the governmental bodies as well as industry to take decisions for the benefit of general public;
- The potential risks and risk management of nanotechnology.

16.9   Conclusions and Future Perspectives

Nanotechnology has come up with many solutions to previously unsolved pharmaceutical, medical and technical problems. It has revolutionised healthcare with its contribution to the betterment of biomarkers, imaging, drug discovery, development and delivery, etc. The applications of nanotechnology in healthcare thus have been growing exponentially, along with increasing interests in investment from both government and industry [35].

However, in healthcare, the unique and novel properties of nanomaterials could become a particular issue in regulatory department due to insufficient information of their toxicity profile and being very different to regular pharmaceuticals. This has caused dilemma among expertise in regulatory body to categorise the nano-related products before getting evaluated for the market approval. The line among many categories are merged and blurred when coming to evaluating nanopharmaceutical. The knowledge gaps on the subject and lack of expert in the field employed among regulation institutes has also created numerous obstacles.

It has also been shown that the public awareness and understanding of nanotechnology in healthcare is still immature and sometimes may be biased and prejudiced due to the concerns of nanotoxicity. Meanwhile, the use of nanotechnology may be promoted too extensively in a sense that it becomes a hype far detached from the reality [36].

Despite such dilemmas, there is no doubt that the future of health will be closely interlinked with developments in nanotechnology which is being used in an evolutionary manner to improve and/or replace many existing therapeutics and healthcare products. It shouldn't be a surprise that in the near future we will have smart 'nanobots' which could be safely be taken by the human body and then automatically repair or destroy specific diseased cells/tumours.

References

1. Tibbals HF (2011) Medical nanotechnology and nanomedicine. CRC Press, Abingdon
2. Farokhzad OC, Langer R (2009) Impact of nanotechnology on drug delivery. ACS Nano 3(1):16–20
3. Booker R, Boysen E (2005) Nanotechnology for dummies. Wiley, Indiana
4. Bhowmik D, Chiranjib CR, Tripathi KK, Kumar KS (2010) Nanomedicine-an overview. Int J PharmTech Res 2(4):2143–2151
5. NHS (2013) About the National Health Service (NHS) [Online] Available at: http://www.nhs.uk/NHSEngland/thenhs/about/Pages/overview.aspx. Accessed 09 June 2014
6. DH statistics release and commentary (2012) NHS imaging and radiodiagnostic activity in England Health. Department of Health, UK. Available at: http://webarchive.nationalarchives.gov.uk/20130402145952/http://transparency.dh.gov.uk/2012/07/10/annual-diagnostics-data/. Accessed 09 June 2014
7. Chakraborty M, Jain S, Rani V (2011) Nanotechnology: Emerging Tool for Diagnostics and therapeutics. http://www.ncbi.nlm.nih.gov/pubmed/21847590. Applied biochemistry and biotechnology. Appl Biochem Biotechnol 165(5–6):1178–1187

8. Bogart LK, Genevieve P, Catherine JM, Victor P, Teresa P, Daniel R, Dan P, & Raphaël L (2014) Nanoparticles for Imaging, Sensing, and Therapeutic Intervention. ACS nano 8(4):3107–3122. http://pubs.acs.org/doi/abs/10.1021/nn500962q

9. Lee DE, Koo H, Sun IC, Ryu JH, Kim K, Kwon IC (2012) Multifunctional nanoparticles for multimodal imaging and theragnosis. Chem Soc Rev 41(7):2656–2672

10. Bao G, Mitragotri S, Tong S (2013) Multifunctional nanoparticles for drug delivery and molecular imaging. Annu Rev Biomed Eng 15:253–282

11. Shin SJ, Beech JR, Kelly KA (2013) Targeted nanoparticles in imaging: paving the way for personalized medicine in the battle against cancer. Integr Biol 5(1):29–42

12. Choi S, Jung GB, Kim KS, Lee GJ, Park HK (2014) Medical applications of atomic force microscopy and Raman spectroscopy. J Nanosci Nanotechnol 14(1):71–97

13. Park K (2013) Facing the truth about nanotechnology in drug delivery. ACS Nano 7(9):7442–7447

14. Safari J, Zarnegar Z (2014) Advanced drug delivery systems: nanotechnology of health design a review. J Saudi Chem Soc 18(2):85–99

15. Junghanns JUA, Müller RH (2008) Nanocrystal technology, drug delivery and clinical applications. Int J Nanomedicine 3(3):295

16. Suzuki Y, Tanihara M, Nishimura Y, Suzuki K, Kakimaru Y, Shimizu Y (1998) A new drug delivery system with controlled release of antibiotic only in the presence of infection. J Biomed Mater Res 42(1):112–116

17. Freitas RA Jr (2005) What is nanomedicine? Nanomedicine: nanotechnology. Biol Med 1(1):2–9

18. Kirtane AR, Panyam J (2013) Polymer nanoparticles: weighing up gene delivery. Nat Nanotechnol 8(11):805–806

19. Gupta A, Arora A, Menakshi A, Sehgal A, Sehgal R (2012) Nanotechnology and its applications in drug delivery: a review. Webmed Central Int J Med Mol Med 3(1), WMC002867

20. Jin S, Leach JC, Ye K (2009) Nanoparticle-mediated gene delivery. http://www.ncbi.nlm.nih.gov/pubmed/19488722. Methods in molecular biology (Clifton, NJ). Methods Mol Biol 544:547–557

21. Donaldson K (2004) Nanotoxicology. Occup Environ Med 61:727–728

22. Borm PJ, Robbins D, Haubold S, Kuhlbusch T, Fissan H, Donaldson K et al (2006) The potential risks of nanomaterials: a review carried out for ECETOC. Part Fibre Toxicol 3(1):11

23. Buzea. http://link.springer.com/search?facet-author=%22Cristina+Buzea%22. Pacheco II C, Robbie K (2007) Nanomaterials and nanoparticles: Sources and toxicity. http://link.springer.com/journal/13758. Biointerphases 2(4):MR17–MR71

24. Ray PC, Yu H, Fu PP (2009) Toxicity and environmental risks of nanomaterials: challenges and future needs. J Environ Sci Health C 27(1):1–35

25. Yah CS, Simate GS, Iyuke SE (2012) Nanoparticles toxicity and their routes of exposures. Pak J Pharm Sci 25(2):477–491

26. Love SA, Maurer-Jones MA, Thompson JW, Lin YS, Haynes CL (2012) Assessing nanoparticle toxicity. Annu Rev Anal Chem 5:181–205

27. HSE (2013) Understanding the hazards of nanomaterials. Health and Safety Executive, UK. [Online] Available at: http://www.hse.gov.uk/nanotechnology/understanding-hazards-nanomaterials.htm. Accessed 24 June 2014

28. Bawa R, Melethil S, Simmons WJ, Harris D (2008) Nanopharmaceuticals: patenting issues and FDA regulatory challenges. The SciTech Lawyer 5(2):1–6

29. Bawarski WE, Chidlowsky E, Bharali DJ, Mousa SA (2008) Emerging nanopharmaceuticals. Nanomed Nanotechnol Biol Med 4(4):273–282

30. Dorbeck-Jung BR, Chowdhury N (2011) Is the European medical products authorisation regulation equipped to cope with the challenges of nanomedicines? Law Policy 33(2):276–303

31. Hullmann A (2006) The economic development of nanotechnology – An indicators based analysis. EU report. http://www.nanotechnology.cz/storage/nanoarticle.pdf. Accessed 25 June 2014

32. Sahoo S, Parveen S, Panda J (2007) The present and future of nanotechnology in human health care. http://www.ncbi.nlm.nih.gov/pubmed/17379166. Nanomedicine: nanotechnology, biology, and medicine. Nanomedicine 3(1):20–31

33. Royal Society & Royal Academy of Engineering (2004) Nanoscience and nanotechnologies: opportunities and uncertainties. https://royalsociety.org/~/media/Royal_Society_Content/policy/publications/2004/9693.pdf. Accessed 09 June 2014

34. Wagner V, Hüsing B, Gaisser S, Bock AK (2008) Nanomedicine: drivers for development and possible impacts. JRC-IPTS, EUR, 23494

35. Schulte J (ed) (2005) Nanotechnology global strategies, industry trend and applications. Wiley, Sussex

36. Timmermans J, Zhao Y. http://www.ncbi.nlm.nih.gov/pubmed?term=van%20den%20Hoven%20J%5BAuthor%5D&cauthor=true&cauthor_uid=22247745. van den Hoven J (2011) Ethics and Nanopharmacy: Value Sensitive Design of New drugs. Nanoethics 5(3):269–283

Chapter 17
The Impact of Nanopharmaceuticals on Healthcare and Regulation

Rebecca Zhangqiuzi Fan, Dan Fei, Roberta D'Aurelio, and Yi Ge

17.1 Introduction

The world is coming to a new era of nanoscience, with various nanotech-based commercial products mushrooming in every sector of the market [1]. For example, the rapid development of nanotechnology offers a broad selection of nanomaterials with precisely controlled manufacture techniques that provide unprecedented advantages in various medical applications [2]. The integration of nanotechnology into medical field, termed as nanomedicine, has been on the European and the U.S. markets for almost two decades (Dorbeck-Jung et al. 2011). Considered as a revolutionary change to the future of medicine, nanomedicine has been developed into a billion-dollar industry enjoying an everlasting rapid boost [3]. Its worth is estimated to reach around $131 billion by 2016 [4].

The appearance of nanotechnology in pharmaceutical world has provided great potentials for the improvement of current medical service. Nanopharmaceuticals already has shown some superior performance over conventional ones with better biocompatibility, bioavailability, and system stability. Furthermore, pharmaceuticals developed at the nano-scale could induce different pharmacokinetics, and have better

R.Z. Fan
Department of Clinical Neurobiology and Institute of Translational
and Stratified Medicine, Plymouth University, Plymouth PL6 8BU, UK

D. Fei
Leicester School of Pharmacy, De Montfort University,
The Gateway, Leicester LE1 9BH, UK

R. D'Aurelio • Y. Ge (✉)
Centre for Biomedical Engineering, Cranfield University, Cranfield, Bedfordshire,
MK43 0AL, UK
e-mail: y.ge@cranfield.ac.uk

© Springer Science+Business Media New York 2014
Y. Ge et al. (eds.), *Nanomedicine*, Nanostructure Science and Technology,
DOI 10.1007/978-1-4614-2140-5_17

solubility, efficacy circulation half-life and lower toxicity with decreased dosage. These advantages would thus significantly benefit patients, who suffer from uncomfortable drug administration and drug resistance. Furthermore, with a smart structure design and functionalisation at nano-scale, clinical therapeutics such as drug delivery could be remarkably enhanced by specific targeting and better penetration, precisely detecting diseases at very early stages before situation deteriorates [3].

On the other hand, the rapid development of science and technology brings about higher requirements for survival and quality of life. Traditional regime is not capable of fulfilling the emerging need for personalized medicine [5]. As the most regulated sector in industry, pharmaceutical manufacture faces risky hedges in marketing and compensation [6]. Although current pharmaceutical regime in the EU is a well-established, flexible and comprehensive system, its appropriateness toward the regulation of nanomedicine has been questioned. In the late 1950s, drug disasters such as thalidomide crisis triggered the launch of new regulations for medicine in Europe to ensure the safety as well as efficiency and quality (Dorbeck-Jung et al. 2011). It is worth noting that nano-based drugs are regulated as conventional products in the process of commercialization at the moment (Dorbeck-Jung et al. 2011). In fact, current guidelines contain no specific standard for nanoparticles, despite more issues (e.g. toxicity) have been generated for their medicinal applications. Thus, specific regulations for the assessment and approval of nanomedicine (including nanopharmaceuticals) products are urged [7].

Overall speaking, nanopharmaceuticals, as a part of nanomedicine engaged in therapeutics, is regarded as one of the most promising subfields of nanomedicine (Dorbeck-Jung et al. 2011). It has generated numerous and increasing interests in the past decade, filling in the communication gaps of nanotechnology, material science biomedical science and pharmaceutical science. Since nanopharmaceuticals could greatly alter the some nature of existing pharmaceuticals (e.g. way of administrating therapeutic agents) [8], inevitably, new considerations for healthcare and corresponding regulations have been raised simultaneously.

17.2 Nanopharmaceuticals in Healthcare

17.2.1 Techniques and Advances

The molecular-levelled interactions of nanomaterials with targets could moderate cellular microenvironment at specific spot, providing a new approach to therapeutics [2]. The diversity of nano-scale materials (Fig. 17.1) provides more options for various applications of nanomedicine, such as disease diagnose and drug delivery. Some commonly used nanopharmaceutical materials are particularly listed in Table 17.1, together with their advantages and pharmaceutical applications.

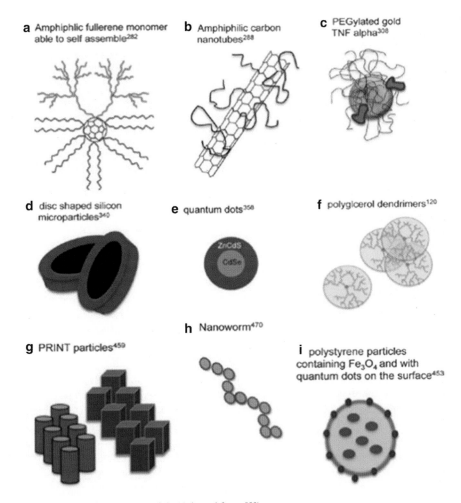

a Amphiphlic fullerene monomer able to self assemble[282]

b Amphiphilic carbon nanotubes[288]

c PEGylated gold TNF alpha[308]

d disc shaped silicon microparticles[340]

e quantum dots[358]

f polyglcerol dendrimers[120]

g PRINT particles[459]

h Nanoworm[470]

i polystyrene particles containing Fe_3O_4 and with quantum dots on the surface[453]

Fig. 17.1 Emerging nanomaterials (Adapted from [9])

As further summarized by Mansour et al. [11] and Hossain Saad et al. [12] recently, nanopharmaceuticals could bring the following advantages:

- Nano size, which increase the surface area, thus enhancing the dissolution rate.
- Surface charge
- Improve drug targeting ability
- Increase the drug stability and improve formulation
- Reduce the dose needed and toxicity related to the drug molecules
- Enhance solubility and bioavailability
- Increase patient compliance since drugs can be deliver by different routes (oral, topical, intravenous, intranasal, etc.)

Table 17.1 Common nanomaterials used in pharmaceutical applications (Modified from [10])

Nanomaterial	Advantage	Application
Carbon nanotubes	Unique thermal, spectroscopic, electrochemical properties	Drug carrier, imaging agents, hybrid theranostics
Liposomes	Low toxicity composition, can incorporate both hydrophilic and lipophilic drugs	Drug delivery, triggered drug release, develop novel probe
Polymer conjugates/micelles	Biodegradable & biocompatible	Drug delivery for a wide range of disease treatment, especially require chronic administration ones
Dendrimers	Capable of penetrating biological barriers, non-immunogenic, surface chemistry is tailored	Drug delivery, MRI imaging, gene therapy, receptor-mediated / passive tumour targeting
Silver nanoparticles	Strong antibacterial action	Aid wound healing, treating atopic dermatitis

The nanotech-incorporated cancer therapeutics is among the top therapeutic/ pharmaceutical applications of nanotechnology [13, 14]. As one of the leading causes of death worldwide without efficient treatment, cancer poses a great threat toward our modern society. The conventional chemotherapy and/or drug treatment could result in severe side effects by non-selectively attacking both healthy and cancer cells. As a solution, nanotechnology could enable a cell-targeting therapy, preventing healthy cells from attack. The unique properties possessed by nanoparticles could also enhance the efficiency of delivery, increase the payload of the drug and further monitor the therapeutic performance [15]. In fact, the second generation of anti-cancer nanomedicine products have started to demonstrate their merits to more stakeholders [8].

Some pH-sensitive nanocarrier systems, for instance, have been successfully developed for targeted therapy in cancer treatment (Manchun et al. 2011). Being different from pH 7.4 for normal tissues, the extracellular pH value of cancer tissues is 6.8 due to the increased extracellular lactate and protons in the microenvironment led by the up-regulated glycolysis. In such systems, the loaded drugs could retain within the nanocarriers at pH 7.4 for normal tissues and be specifically released at an acidic condition near the cancer tissues as a result of structure and/or formation change of the nanocarriers. For example, Filippov and his colleagues [16] recently reported a HPMA (N-(2-hydroxypropyl)-methacrylamide)-based nanoparticle-drug conjugate for targeted drug delivery. The drug doxorubicin (Dox) was conjugated to the nanoparticle via a pH-responsive hydrazine bond. Moreover, the size and shape along with internal structure of the conjugate could be monitored and controlled precisely.

Apart from anticancer therapy, nanopharmaceuticals also has successfully provided promising treatments for other clinical conundrums such as Alzheimer's disease [17], diabetes [18], and ocular delivery (Vadlapudi et al. 2013).

In terms of the fabrication of nanoparticles in pharmaceutical manufacturing, various techniques have been engaged, each with its own advantages and drawbacks [19]:

- Milling: Traditional milling techniques can be modified to operate under certain speed to reduce the particle size to nano size. This technique faces the challenges of balancing the type and amount of the additives (influence toxicity) with the stability of the system, purification of the product, and the variety of time consumed for an aimed range of sizes.
- Polymerisation: This is a widely adapted bottom-up technique for fabricating nanoparticles by using monomers as starting materials for polymerisation. The main disadvantage of this technique is the (potential) toxicity of the monomers and resulting products.
- Emulsification (/Precipitation/Coacervation): In this technique, the drug is dissolved and mixed in an organic, miscible anti-solvent to form a stable emulsion. The product is received after removing the solvent. The drawbacks of this technique are variable mixing processes resulting in various range of size distribution of the nanoparticles, and the spontaneous growth of crystals in presence of a nucleation that increases the difficulty to control the size distribution. Also, the removal of solvent must be sufficient; otherwise the residue of it in the nanoparticles may cause the degradation of nanoparticles.
- Microfluidisation: Also termed as piston-gap homogenizer, this technique prepares nanoparticles/nanosuspensions via homogenization under a high pressure. Problems for this technique include requiring larger amounts of energy for further reduction in size, blocking of the piston, the difficulty for scale-up, and contamination by heavy metal in some cases.
- Supercritical Fluid Technique: This is an environment-friendly technique manufacturing nanoparticles via a procedure free from solvent. It has become an alternative for producing nanoparticles in recent years on the basis of that of micro-scale particles. The restriction of its utility is the limited choices for the polymer since only few polymers have shown to be soluble in supercritical fluid.
- Spray drying: This method fabricates nanoparticles by spray drying of the suspension of nanoparticles, which is prepared through wet comminution using stabilizers.

17.2.2   Positive Impacts

Nanotechnology has already demonstrated its power and capability in revolutionising many healthcare practice and treatments, such as cancer and cardiovascular diseases treatment, gene therapy, and orthopaedic implants. For pharmaceuticals, the unique properties and advantages of nanomaterials have remarkably affected and enhanced the whole production line of this business, from drug discovery and development to drug delivery. As a result, great positive impacts have been made on healthcare, in terms of higher efficiency of treatment, decreased healthcare cost, and less patient suffering.

17.2.3   Safety and Other Concerns

Apart from the great potentials, nanopharmaceuticals has to be carefully examined against uncertainties, such as knowledge gaps on nanomaterials' physicochemical properties and toxicity aspects. The detection and characterisation methods as well as toxicological data of nanopharmaceutical products are still lacking. As a result, the related risk assessment is incomplete. In recent years, increasing concerns have been raised about the long-term potential hazard of employing nanomaterials, resulting in possible toxicity (e.g. cytotoxicity and genotoxicity). As it is well known that the penetration of conventionally impermeable biological barriers can sometimes trigger neurotoxicity, carbon nanotubes, for instance, have been reported to have similar pathogenic phenomenon as asbestos in mice [20]. Intratracheal administration of single-walled carbon nanotubes has also been proven to exacerbate allergen-related airway inflammation (Syed et al. 2012). The specificity of nanomaterials undoubtedly leads to a necessity of the multidisciplinary investigation of nanopharmaceuticals that involves various stages [21]. Unfortunately, the progress is hampered by the unclear and yet-to-reveal hidden mechanism of interactions between nanomaterials and biosystems (e.g. cells, tissues, and organs).

Because of the vital role of pharmaceutical products plays in the battle of saving life, safety needs to be considered as a priority in assessing a new nanopharmaceutical product. Early assessments of safety therefore shall be the open-shut gate deciding whether the product is worthwhile of a further clinical development (Fig. 17.2) [9].

17.3   Nanopharmaceuticals and Regulation

Nanopharmaceuticals are creating new challenges for the regulatory bodies – such as the Food and Drug Administration (FDA) and the European Medicines Agency (EMA). The first challenge is about a universal and international definition of what is nanotechnology [22]. Without a proper and universally agreed definition, it is difficult to put proper regulations in place that take into account all aspects of the use of nanomaterials in pharmaceuticals. For example, nanomaterials could be defined as materials with a size smaller than 100 nm, while some others accept larger materials with a size up to 500 nm as nanomaterials [23].

Pharmaceutical industry is one of the most regulated and controlled industrial sectors. There are already many commercially available nanopharmaceutical products (e.g. Myocet and Rapamune). Most of them have been approved by the regulatory body based on the pre-existing laws/regulations without a further special assessment. Due to the amazing development of nanopharmaceuticals in recent years and its safety concerns discussed earlier in the chapter, the new/refined regulation is urged with a special assessment both in vivo and in vitro.

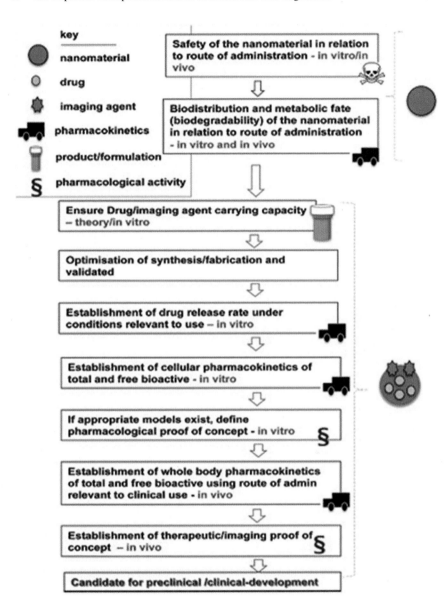

Fig. 17.2 Summary of stop-go checkpoints for the preclinical development of nanomedicine (Adapted from [9])

Excellent progress has been made. Under FDA's updated guidance, the following points should be considered by industry, when a product involves the application of nanotechnology or contains nanomaterials [24]:

- Whether a material or end product is engineered to have at least one external dimension, or an internal or surface structure, in the nanoscale range (approximately 1–100 nm). They are relevant to evaluations of safety, effectiveness, performance, quality, public health impact, or regulatory status of products;
- Whether a material or end product is engineered to exhibit properties or phenomena, including physical or chemical properties or biological effects, that are attributable to its dimension(s), even if these dimensions fall outside the nanoscale range, up to one micrometer (1,000 nm). They emphasized that "such evaluations should include a consideration of the specific tests (whether traditional, modified, or new) that may be needed to determine the physicochemical properties and biological effects of a product that involves the application of nanotechnology".

By contrast, REACH (Registration, Evaluation, Authorisation and Restriction of Chemicals), which is a regulation of the European Union adopted to improve the protection of human health and the environment from the risks that can be posed by chemicals, entered into force on 1 June 2007. Based on the communication of second regulatory review on nanomaterials in 2012, a study has recently be undertaken to support the impact assessment of relevant regulatory options for nanomaterials in the framework of REACH [25]. It was suggested that "those nanomaterials currently falling outside existing notification, registration or authorisation schemes" would be considered and included.

17.4 Summary and Future Outlook

Nanotechnology is revolutionising the market of pharmaceuticals [26]. In general, the expanding application of nanopharmaceuticals in medical service is changing our healthcare system as well as the regulatory system.

However, as discussed earlier, there are still many challenges for nanopharmaceuticals with respect to its application and fate in healthcare and regulation, respectively:

- Apart from the challenges of fabrication of nanomaterials used for nanopharmaceuticals, there are also challenges on the more comprehensive characterisation of nanomaterials such as the determination of structure, aggregation, and purity etc. In addition, the safety assessment of nanoproducts is a key for their applications in pharmaceuticals [19]. By considering the large investment and long period of product development in pharmaceuticals, the prediction and/or initial pre-assessment about the biological performance and toxicity outcome of nano pharmaceutical products would become significantly valuable and rewarding.

- Despite the great efforts made on regulation from FDA and the European Commission etc., so far there has been no thorough regulatory system specific for nanopharmaceuticals. The challenges met to obtain sufficient information about the physicochemical properties of nanomaterials, and to understand the mechanism of their molecular and cellular interactions with the internal environment, would underlay and catalyse the formation of a unified regulatory system. At the moment, the most direct challenge for nanopharmaceuticals is to meet all safety requirements in the guidelines set by the regulatory bodies, in order to gain the entrance for clinical use [27].

Inspired by the great impacts of nanopharmaceuticals on both healthcare and regulations, it is reasonable to believe that nanopharmaceuticals holds great potential to form the mainstream of pharmaceutical industry in the future.

References

1. Rivera Gil P, Hühn D, del Mercato LL et al (2010) Nanopharmacy: inorganic nanoscale devices as vectors and active compounds. Pharmacol Res 62(2):115–125
2. Zhang X, Xu X, Bertrand N et al (2012) Interactions of nanomaterials and biological systems: implications to personalized nanomedicine. Adv Drug Deliv Rev 64(13):1363–1384
3. Bawarski WE, Chidlowsky E, Bharali DJ et al (2008) Emerging nanopharmaceuticals. Nanomed Nanotechnol Biol Med 4(4):273–282
4. Bhargavi C, Anil DB, Praneta Desale K (2013) Nanotherapeutics – an era of drug delivery system in nanoscience. Indian J Res Pharm Biotechnol 1(2):210–214
5. Steinbach OC (2013) Industry update: the latest developments in therapeutic delivery. Ther Deliv 4(1):17–20
6. Eaton MAW (2011) How do we develop nanopharmaceuticals under open innovation? Nanomed Nanotechnol Biol Med 7(4):371–375
7. Gaspar R (2007) Regulatory issues surrounding nanomedicines: setting the scene for the next generation of nanopharmaceuticals. Nanomedicine 2(2):143–147
8. Moghimi SM, Peer D, Langer R (2011) Reshaping the future of nanopharmaceuticals: ad iudicium. ACS Nano 5(11):8454–8458
9. Duncan R, Gaspar R (2011) Nanomedicine(s) under the microscope. Mol Pharm 8(6): 2101–2141
10. Duncan R (2011) Polymer therapeutics as nanomedicines: new perspectives. Curr Opin Biotechnol 22(4):492–501
11. Mansour HM, Park C, Bawa R (2014) Design and development of approved nanopharmaceutical products. In: Bawa R, Audette GF, Rubinstein I (eds) Handbook of clinical nanomedicine: from bench to bedside, 1st edn. Pan Stanford Publishing/Taylor & Francis, Singapore, pp 1–27
12. Hossain Saad MZ, Jahan R, Bagul U (2012) Nanopharmaceuticals: a new perspective of drug delivery system. Asian J Biomed Pharm Sci 2(14):11–20
13. Syed S, Zubair A, Frieri M (2013) Immune response to nanomaterials: implications for medicine and literature review. Curr Allergy Asthma Rep 13(1):50–57. http://www.ncbi.nlm.nih.gov/pubmed/22941559
14. Yu MK, Park J, Jon S (2012) Magnetic nanoparticles and their applications in image-guided drug delivery. Drug Deliv Transl Res 2(1):3–21
15. Farrell D, Ptak K, Panaro NJ et al (2011) Nanotechnology-based cancer therapeutics – promise and challenge – lessons learned through the NCI alliance for nanotechnology in cancer. Pharmacol Res 28(2):273–278

16. Filippov SK, Chytil P, Konarev PV et al (2012) Macromolecular HPMA-based nanoparticles with cholesterol for solid-tumor targeting: detailed study of the inner structure of a highly efficient drug delivery system. Biomacromolecules 13(8):2594–2604

17. Ismail MF, Elmeshad AN, Salem NA (2013) Potential therapeutic effect of nanobased formulation of rivastigmine on rat model of Alzheimer's disease. Int J Nanomed 8:393–406

18. Krol S, Ellis-Behnke R, Marchetti P (2012) Nanomedicine for treatment of diabetes in an aging population: state-of-the-art and future developments. Nanomed Nanotechnol Biol Med 8:S69–S76

19. Shah RB, & Mansoor AK (2009) "Nanopharmaceuticals: Challenges and regulatory perspective." In Nanotechnology in Drug Delivery, (pp. 621–646). Springer New York. http://link.springer.com/chapter/10.1007/978-0-387-77668-2_21

20. Kolosnjaj J, Henri S, & Fathi M (2007) Toxicity studies of carbon nanotubes." In Bio-Applications of Nanoparticles, (pp. 181–204). Springer New York. http://www.ncbi.nlm.nih.gov/pubmed/18217344

21. Domingo C, Saurina J (2012) An overview of the analytical characterization of nanostructured drug delivery systems: towards green and sustainable pharmaceuticals: a review. Anal Chim Acta 744:8–22. http://www.ncbi.nlm.nih.gov/pubmed/22935368

22. Bawa R (2013) FDA and nanotech: baby steps lead to regulatory uncertainty. In: Bagchi D, Bagchi M, Moriyama H, Shahidi F (eds) Bio-nanotechnology: a revolution in food, biomedical and health sciences, 1st edn. Wiley, Oxford, pp 720–732

23. Bawa R (2011) Regulating nanomedicine – can the FDA handle it? Curr Drug Deliv 8(3):227–234

24. FDA (2014) Considering whether an FDA-regulated product involves the application of nano-technology. Available http://www.fda.gov/regulatoryinformation/guidances/ucm257698.htm#points. Accessed 18 July 2014

25. Matrix Insight Ltd (2014) Request for services in the context of the FC ENTR/2008/006, lot 3: a Study to support the impact assessment of relevant regulatory options for nanomaterials in the framework of REACH. Available http://ec.europa.eu/DocsRoom/documents/5826/attachments/1/translations/en/renditions/native. Accessed 18 July 2014

26. Bhogal N (2009) Regulatory and scientific barriers to the safety evaluation of medical nano-technologies. Nanomedicine 4(5):495–498

27. Wang R, Billone PS, Mullett WM (2013) Nanomedicine in action: an overview of cancer nanomedicine on the market and in clinical trials. J Nanomater. Article ID:629681

28. Dorbeck-Jung BR, Chowdhury N (2011) Is the European medical products authorisation regulation equipped to cope with the challenges of nanomedicines? Law Policy 33(2):276–303. http://doc.utwente.nl/78741/

29. Manchun S, Dass CR, Sriamornsak P (2012) Targeted therapy for cancer using pH-responsive nanocarrier systems. Life Sci 90(11–12):381–387. http://www.ncbi.nlm.nih.gov/pubmed/22326503

30. Vadlapudi AD, Mitra AK (2013) Nanomicelles: an emerging platform for drug delivery to the eye. Ther Deliv 4(1):1–3. http://www.ncbi.nlm.nih.gov/pubmed/23323774

Chapter 18
Nanomedicine in Cancer Diagnosis and Therapy: Converging Medical Technologies Impacting Healthcare

Maya Thanou and Andrew D. Miller

18.1 Introduction

Nowadays cancer diagnosis and therapy is the primary preoccupation of nanomedicine. This focus has given rise to the new field of cancer nanotechnology that involves multidisciplinary, problem driven research cutting across the traditional boundaries of biology, chemistry, engineering and medicine with the aim of creating major advances in cancer detection, diagnosis and treatment [1–4]. The field has received strong support especially in the US where several nanotechnology for cancer centres have been launched and operated since 2004. There is no better definition and overview of this field, than that given in http://nano.cancer.gov/, which outlines the National Cancer Institute's (NCI's) alliance for nanotechnology for cancer. This alliance aims to create a multidisciplinary nanotechnology approach for the creation of solutions for cancer detection, imaging and diagnosis [5]. In Europe a number of academic groups are interested in cancer nanotechnology as well. However only with the advent of Europe FPVII programs have specific calls been announced to support multidisciplinary research in cancer nanotechnology. In the UK, the major cancer research organisation (Cancer Research UK) appears hesitant to support this emerging field, possibly due to the perceived safety risk from nanomaterials currently untested in man. This hesitation is unfortunate. In a recent report "Roadmaps

M. Thanou
Institute of Pharmaceutical Science, Kings College London,
Franklin-Wilkins Building, Waterloo Campus, 150 Stamford Street, London SE1 9NH, UK

A.D. Miller (✉)
Institute of Pharmaceutical Science, Kings College London,
Franklin-Wilkins Building, Waterloo Campus, 150 Stamford Street, London SE1 9NH, UK

GlobalAcorn Ltd, London, UK
e-mail: a.miller07@btinternet.com; andrew.david.miller@kcl.ac.uk;
andrew.miller@globalacorn.com

© Springer Science+Business Media New York 2014
Y. Ge et al. (eds.), *Nanomedicine*, Nanostructure Science and Technology,
DOI 10.1007/978-1-4614-2140-5_18

in Nanomedicine towards 2020" [6], specialists are now predicting that imaging and therapy in oncology by means of cancer nanotechnology will be a primary opportunity for various "designer" type nanomaterials, nanodevices and nanoparticles currently in discovery and development. Indeed, the global market size for cancer nanotechnology products is predicted to be €30bn by 2015. The particular opportunity presented by cancer nanotechnology is the eventual likelihood of personalised cancer diagnosis and treatment regimes [3].

Personalized therapy of cancer begins with molecular profiling. Golub et al. were first to report how molecular profiling studies, that show variations in gene expression patterns with time and disease status, could be used to inform on the stage, grade, clinical course and response to treatment of tumours [7]. From then on, increasing numbers of such studies have been performed showing that any given metastatic lesion results from a corresponding combination of tumoral, stromal, and inflammatory factors [8, 9]. Following this, causality in cancer has become associated with cancer disease-specific biomarkers validated by histochemical studies of diseased tissue [10]. The identification of such biomarkers by molecular profiling provides the foundation for personalized cancer diagnosis and therapy [11, 12]. In a prime early example of this principle, Erbb2 (HER2) is a tyrosine kinase receptor and cancer disease-specific biomarker found in 25–30 % of breast cancers. Overexpressed HER2 can be targeted for breast cancer therapy using Herceptin that is a potent, anti-HER2 therapeutic monoclonal antibody (biopharmaceutical agent). However, Herceptin has significant drug-use side effects that can be very severe. Accordingly, the Federal Drug Administration (FDA) now requires the proven identification of over-expressed HER2 in breast cancer patients before Herceptin can be prescribed. Typical in vitro diagnostic tests for HER2 that may be used to diagnose the presence of HER2 in breast tumour development include an immunohistochemistry assay and a nucleic acid fluorescence in situ hybridisation (FISH) test. Once these tests can be shown positive, then breast cancer patients may then be prescribed Herceptin with real confidence in probable therapeutic outcomes. In summary, cancer disease-specific biomarker, HER2, is detected as a diagnosis for breast cancer and disease mechanism. Afterwards a biomarker selective biopharmaceutical agent can be administered.

Relevant cancer disease-process biomarkers are many and various. They range from mutant genes, non-coding RNAs (ncRNAs), proteins, lipids, to carbohydrates and may even be small metabolite molecules. The key is that a link(s) should be established clearly from a given biomarker to tumour growth and development. Following on from this, there is a definite requirement for hyper-flexible, platform technologies that can mobilize diagnostic agents for a given biomarker and then deliver biomarker selective therapeutic agents to disease-target cells, also with selectivity. From the various options open to cancer nanotechnology, multi-functional nanoparticles are potentially ideal to meet these twin requirements. Indeed nanoparticles could be envisaged for (a) the detection of biomarkers, (b) the imaging of tumours and their metastases, (c) the functional delivery of therapeutic agents to target cells, and (d) the real time monitoring of treatment in progression. Therefore, if this is the potential, how close are we really?

Where nanoparticles are to be created for the functional delivery of imaging and/or therapeutic agents specific to cancer biomarkers, many factors have to be taken into consideration. This fact can be illustrated with reference to the fields of gene therapy and RNA interference (RNAi) therapeutics where nanoparticles have been devised for functional delivery of therapeutic nucleic acids with some success [13–15]. Where nanoparticles have been successfully designed and used to mediate the functional delivery of therapeutic nucleic acids, an **ABCD** nanoparticle paradigm can be invoked (Fig. 18.1). According to this general paradigm, functional nanoparticles comprise active pharmaceutical ingredients (APIs) (**A**-components) surrounded initially by compaction/association agents (**B**-components – typically lipids, amphiphiles, proteins or even synthetic polymers etc.) designed to help sequester, carry and promote functional delivery of the **A**-components. Such core **AB** nanoparticles may have some utility in vivo but more typically require coating with a stealth/biocompatibility polymer layer (**C**-layer; primary **C**-component – most often polyethylene glycol [PEG]) designed to render resulting **ABC** nanoparticles with colloidal stability in biological fluids and with immunoprotection from the reticuloendothelial system (RES) plus other immune system responses. Finally, an optional biological targeting layer (**D**-layer; primary **D**-components – *bona fide* biological receptor-specific targeting ligands) might be added to confer the resulting **ABCD** nanoparticle with target cell specificity. A key design principle here is that tailor-made nanoparticles can self-assemble reliably from tool-kits of

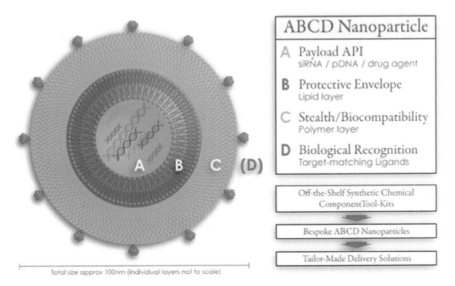

Fig. 18.1 Active pharmaceutical ingredients (APIs) (therapeutic bio-actives or intractactable drugs) are condensed within functional concentric layers of chemical components making up nanoparticle structures designed to enable efficient delivery (trafficking) of active therapeutic agents to disease-target cells. **ABCD** nanoparticle is drawn here assuming that **A**-components are nucleic acids and that **B**-components employed are lipids

purpose designed chemical components [16–26]. Accordingly, the concept of a personalized nanoparticle formulation, assembled in the pharmacy for an individual patient does not seem so far removed from reality.

The **ABCD** nanoparticle paradigm represents a set of well-found principles of design that are being implemented in the real world with the formation of actual nanoparticles leading to actual demonstrated functional properties at least in pre-clinical studies. As such, the design principles laid out in the **ABCD** nanoparticle paradigm are widely corroborated in the literature [1, 27–35]. Clearly functional nanoparticles need to be constructed from a range of chemical components designed to promote functional delivery of different diagnostic and/or therapeutic agents in vivo. In practise this means that nanoparticles need to be equipped to overcome relevant "bio-barriers" in accordance with pharmacological requirements of API use such as site, time and duration of action. Importantly too, with clinical goals in mind, nanoparticles have to be considered different to small and large molecular drugs. For instance, regulations from the FDA state that Absorption, Distribution, Metabolism and Excretion (ADME) studies need to be redesigned in the case of nanoparticles to take into consideration their aggregation and surface chemical characteristics [36].

In terms of cancer diagnosis and therapy, there is one factor that is very much in favour of multifunctional nanoparticle use. Nanoparticles administered in the blood stream (i.v.-administration) frequently accumulate in tumours anyway due to the enhanced permeability and retention (EPR) effect, a behaviour that was identified by Maeda as a means to target anticancer therapeutic agents to tumours [37, 38]. Nanoparticle accumulation in tumours takes place due to the presence of highly permeable blood vessels in tumours with large fenestrations (>100 nm in size), a result of rapid, defective angiogenesis. In addition tumours are characterised by dysfunctional lymphatic drainage that helps the retention of nanoparticles in tumour for long enough to enable local nanoparticle disintegration in the vicinity of tumour cells. The phenomenon has been used widely to explain the efficiency of nanoparticle and macromolecular drug accumulation in tumours [39]. Unfortunately, knowledge of nanoparticle biokinetics, metabolism and clearance is otherwise poor since too few nanoparticle products have been clinically tested. This is a major limitation in the growth of the field of cancer nanotechnology. Nevertheless, cancer nanotechnology is a fast growing field and new data is arriving all the time. In the following sections, the status of nanoparticle use in cancer diagnosis and therapy will be surveyed.

18.2   Nanoparticles for Cancer Imaging and Therapy in Clinical Trials and at Advanced Preclinical Phases of Evaluation

The first nanoparticles used and approved for clinical therapy use were lipid-based nanoparticles (LNPs). Selected structural lipids self-assemble into liposomes that are typically approx. 100 nm in diameter and consist of a lipid bilayer surrounding an aqueous cavity [40–43]. This cavity can be used to entrap water-soluble drugs in an enclosed volume resulting in a drug-**AB** nanoparticle system [44, 45]. The first reported LNPs of this type were designed to improve the pharmacokinetics and biodistribution of the anthracycline drug doxorubicin. Doxorubicin is a potent anti-cancer agent but is cardiotoxic. In order to minimize cardiotoxicity, doxorubicin was encapsulated in anionic liposomes giving anionic doxorubicin drug-**AB** nanoparticles that enabled improved drug accumulation in tumours and increased antitumour activity while diminishing side effects from cardiotoxicity [46, 47]. This nanoparticle formulation has since been used efficiently in clinic for the treatment of ovarian and breast cancer [48, 49]. Thereafter, Doxil® was devised corresponding to a drug-**ABC** nanoparticle system, comprising PEGylated liposomes with encapsulated doxorubicin. These Doxil® drug-**ABC** nanoparticles (also known as PEGylated drug-nanoparticles) were designed to improve drug pharmacokinetics and reduce toxicity further by maximizing RES avoidance [50–52], making use of the PEG layer to reduce uptake by RES macrophages of the mononuclear phagocyte system (MPS) [53, 54].

The second nanoparticle system used and approved for clinical use were nanoparticles prepared using albumin as a compaction/association agent for sparingly water soluble Taxol®, one of the most potent anticancer drugs known. The resulting protein-based drug-**AB** nanoparticles (130 nm diameter) were christened nab-paclitaxel or Abraxane®. This Abraxane® system was designed to avoid the use of Cremophor EL® solvent (polyethoxylated castor oil) most frequently used to solubilise Taxol® [55–57]. Abraxane® is the first albumin nanoparticle system approved for human use by the FDA. This use of albumin is inspired. Albumin is a natural carrier of endogenous hydrophobic molecules that associate through non-covalent interactions. In addition, albumin assists endothelial trancytosis of protein bound and unbound plasma constituents principally through binding to a 60 kDa glycoprotein cell surface receptor, gp60. The receptor then binds to caveolin-1 with subsequent formation of transcytotic vesicles (caveolae) [58]. In addition, albumin binds to osteonectin, a secreted protein acid rich in cysteine (SPARC), that is present on breast lung and prostate cancer cells, so allowing albumin nanoparticles to accumulate readily in tumours [57, 59]. Currently there are more than 50 clinical trials ongoing using nanoparticles for cancer therapy. Indeed, the majority of these nanoparticles are nab-type (nanoparticle albumin bound) tested for the treatment of various cancer types (http://clinicaltrials.gov).

Otherwise, in terms of leading edge cancer clinical trials, LNPs have also been used in clinical trials for the delivery of biotherapeutic agents in cancer therapy corresponding to leading RNAi effectors known as small interfering RNAs (siRNAs). For instance LNPs corresponding to siRNA-**ABC** nanoparticles, Atu027, ALN-VSP02 and TKM-PLK1 are or have been in various stages of Phase I clinical trials. Moreover, one polymer-based nanoparticle (PNP) system, corresponding to a siRNA-**ABCD** nanoparticle system and christened CALAA-01, has appeared in Phase I clinical trials, with a Phase IIa clinical trial reportedly underway [60]. CALAA-01 employs a cyclodextrin polymer scaffold to entrap RNAi effectors and transferrin as a receptor-specific targeting ligand. Otherwise, advanced LNP (and even PNP) prototypes, that are either nucleic acid-**AB**, **ABC** or **ABCD** nanoparticle systems, continue to be tested for functional delivery of therapeutic nucleic acids to target cells in animal models of human disease (to liver for treatment of hepatitis B and C virus infection, to ovarian cancer lesions for cancer therapy) and to target cells in murine lungs [61–67]. Rules for enhancing efficient delivery through receptor-mediated uptake of **ABCD** nanoparticles into target cells are also being studied and appreciated [68–71].

From the point of view of using nanoparticle technologies for the imaging of cancer, the ability to combine imaging agents with nanoparticles is central. In terms of the **ABCD** nanoparticle paradigm, the **A**-component now becomes an imaging agent(s) instead of a therapeutic agent. Fortunately, progress with imaging nanoparticles has also been brisk and a number of clinical trials have been expedited. For instance, a heterogeneous LNP system has been described in clinic that consists of a superparamagnetic iron oxide (SPIO) core particle lipid-coated to confer biological function [72]. This LNP system been used as a diagnostic tool for the pre-operative stage(s) of pancreatic cancer [73]. LNPs have also been described for radionuclide delivery to tumour lesions. Typically, these consist of a central liposome, that entraps a radionuclide of interest by analogy to drug-**AB/C** nanoparticles, and whose surface may be modified by targeting antibodies or peptides (**D**-components) in order to derive receptor-targeted nanoparticles [74]. Nanoparticles of this type have been used to entrap the chelate ^{111}In-diethylenetriamine-pentaacetic acid (^{111}In-DTPA). These were administered to 17 patients with locally advanced cancers. Post administration, patients were examined by means of a whole body gamma camera in order to verify pharmacokinetics and biodistribution behaviour. The $t_{1/2}$ of these ^{111}In-labelled nanoparticles was 76.1 h, and levels of tumour LNP uptake were estimated to be approximately 0.5–3.5 % of the injected dose at 72 h. The greatest levels of uptake were seen in the patients with head and neck cancers. However, significant uptake was also seen in the tissues of the RES (namely, liver, spleen, and bone marrow). Nevertheless data support the use of these ^{111}In-labelled nanoparticles for the imaging of solid tumors, particularly those of the head and neck, [75]. Moreover, once delivered to such tumour lesions, the radionuclide may then be used as a therapeutic agent to destroy tumour mass by radiation according to the principles of nuclear medicine.

Potentially important preclinical studies have been carried out recently with imaging LNPs set up for positive contrast magnetic resonance imaging (MRI) [76, 77]. The first described LNPs of this class were formulated by trapping water-soluble, paramagnetic, positive contrast imaging agents [such as $MnCl_2$, gadolinium (III) diethylenetriamine pentaacetic acid (Gd.DTPA), and the manganese (II) equivalent (Mn.DTPA)] in the enclosed volume of a liposome resulting in prototype lipid-based, positive contrast imaging-**AB/C** nanoparticles [78, 79]. Disadvantages were quickly reported such as poor encapsulation efficiency, poor stability, and clear toxicities due to importune contrast agent leakage and poor relaxivity [80]. These problems were obviated when hydrophobic lipidic chains were "grafted" on to contrast agents, thereby enabling these agents to be hosted by a lipid-bilayer [81]. Such lipidic contrast agents formulated in association with the bilayer of a liposome exhibit improved ionic relaxivity and could therefore be used for dynamic MRI experiments in mice in vivo [82].

A potentially significant variation on this theme involves gadolinium (III) ions complexed with 1,4,7,10-tetraazacyclododecane-1,4,7,10-tetraacetic acid (DOTA) to which hydrophobic lipidic chains are attached. In particular, gadolinium (III) 2-(4,7-*bis*-carboxymethyl-10-[(N,N-distearylamidomethyl)-N'-amidomethyl]-1,4,7,10-tetraazacyclododec-1-yl)-acetic acid (Gd.DOTA.DSA) was prepared and formulated into passively targeted Gd-**ABC** (no biological targeting layer) and folate-receptor targeted Gd-**ABCD** nanoparticles in conjunction with a number of other naturally available and synthetic lipid components such as (ω-methoxy-polyethylene glycol 2000)-N-carboxy-distearoyl-L-α-phosphatidylethanolamine (PEG^{2000}-DSPE) or its folate variant (folate-PEG^{2000}-DSPE), and fluorescent lipid dioleoyl-L-α-phosphatidylethanolamine-N-(lissamine rhodamine B sulphonyl) (DOPE-Rhoda) (Fig. 18.2). These bimodal imaging nanoparticle systems demonstrated excellent tumour tissue penetration and tumour MRI contrast imaging in both instances [83–85]. Interestingly, the folate-receptor targeted Gd-**ABCD** exhibited a fourfold decrease in tumor T_1 value in just 2 h post-injection, a level of tissue relaxation change that was observed only 24 h post administration of passively targeted Gd-**ABC** nanoparticles [83, 84]. Preparations for clinical trial are now underway beginning with cGMP manufacturing and preclinical toxicology testing. These Gd-**ABC/D** nanoparticles are potentially excellent nanotechnology tools for the early detection and diagnosis of primary and metastatic cancer lesions. How effective remains to be seen when clinical trials can be performed. On the other hand, these LNPs may well enter into direct comparison with alternative LNPs that have been described by Müller et al. and are known as solid lipid nanoparticles (SLNs). These SLNs could certainly offer an alternative LNP platform for imaging [86–88]. For instance, under appropriate optimised conditions SLNs can carry MRI contrast agents [89], and SLNs containing [Gd-DTPA(H$_2$O)]$^{2-}$ and [Gd-DOTA(H$_2$O)]$^-$ have even been prepared for preclinical studies.

In complete contrast, a variety of PNP systems are also beginning to be realized for the delivery of therapeutic agents and/or imaging agents. For instance, dendrimers are a unique class of repeatedly branched polymeric macromolecules with a nearly perfect 3D geometric pattern. They can be prepared with either divergent methods (outward from the core) or convergent methods (inward towards the core). Tomalia was the first to synthesise the 3D polyamidoamine (PAMAM) dendrimers using divergent methods [90]. The methods of Frechet [91] are characterised by generation (G) building using monomers added to a central core. Controlled synthesis results in molecular diameters between 1.9 nm for G1 to 4.4 nm for G4 dendrimers. These G1-G4 dendrimers represent the smallest known nanocarriers yet developed for pharmaceutical and imaging applications associated with cancer [92], including photodynamic therapy (activation therapies) [93], boron neutron capture therapy [94] and hyperthermia therapies in combination with gold nanoparticles [3]. These Gd-**AB** nanoparticles, known as gadolinium (III) dendrimer conjugates, have proven of provisional value in MRI experiments [95]. Unfortunately as delivery systems for therapeutic agents, dendrimers have a tendency post administration to release conjugated drugs before reaching disease target sites.

a   Gd-ABC/D nanoparticles; *in vivo* delivery

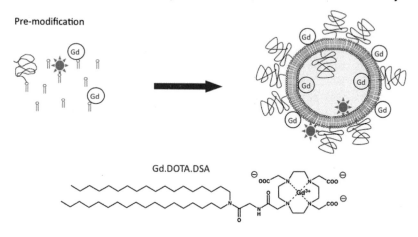

Pre-modification

Gd.DOTA.DSA

LTC Gd-ABC nanoparticles (Gadonano) for MRI and fluorescence imaging

Gd.DOTA.DSA/DOPC/Chol/DSPE-PEG^{2000}/DOPE-Rhodamine (30:32:30:7:1, m/m/m/m/m)
size: ~ 100 nm (PCS and cryo-EM); net charge ~ neutral

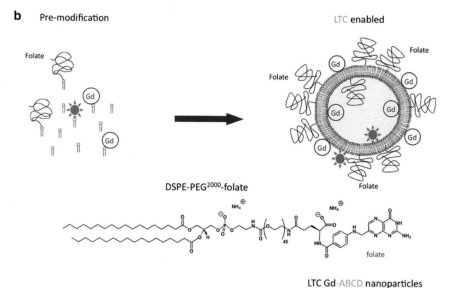

b   Pre-modification   LTC enabled

Folate

DSPE-PEG^{2000}-folate

folate

LTC Gd-ABCD nanoparticles

Fig. 18.2 Passively targeted Gd-**ABC** (*top*) and folate-receptor targeted Gd-**ABCD** (*bottom*) nanoparticles for IGROV-1 tumour imaging [83]. These LNPs are long-term circulation (*LTC*) enabled by virtue of the use of bilayer stabilizing lipids and 7 mol% PEG-lipid in the outer leaflet membranes of lipid-based nanoparticle structures

Finally, we turn to inorganic "hard" nanoparticles. Of these the most advanced already in clinical practice are the dextran coated iron oxide nanoparticles that correspond in form to imaging-**AB/C** nanoparticle systems. Ferumoxtran-10 ® is a commercially available ultra-small-superparamagnetic iron oxide particle (USPIO) product [96, 97]. After systemic injection, these nanoparticles collect in lymph nodes, liver, spleen, or brain tissue where are visualized by MRI. In a lymph node with proper architecture and function (healthy), macrophages take up a substantial amount of ferumoxtran-10. This uptake results in a marked reduction in signal intensity and turns the lymph nodes dark when seen by MRI. Infiltration of lymph nodes with malignant cells replaces the macrophages and changes the architecture of the lymph nodes. In malignant lymph nodes there is no ferumoxtran-10 macrophage uptake and they can retain the high signal intensity or display heterogenous signal intensity if micrometastases are involved. This way the grade of tumours and prognosis can be assessed by the presence of micro-metastases [98]. Additionally, iron oxide nanoparticles can be guided in principle to target sites (i.e. tumour) using external magnetic field and they can be also heated to provide hyperthermia for cancer therapy [99].

On another tack, Yu et al. have reported how dextran-coated iron oxide nanoparticles bearing a Cy5.5 near infrared (NIR) probe could also carry doxorubicin thereby allowing both the imaging and drug treatment of cancer lesions. The administration of these bimodal imaging drug-**AB/C** nanoparticles allowed for simultaneous real-time imaging of nanoparticle biodistribution and the measurement of drug pharmacokinetic behaviour alongside the observation of a substantial phenotypic (pharmacodynamic) reduction in tumour size [99]. Similarly bimodal imaging RNAi-**AB/C** nanoparticle systems were realized by coupling RNAi effectors to the dextran coat alongside Cy5.5 near infrared dye. These bimodal imaging nanoparticles were also seen to enable functional delivery of the RNAi effectors to target cells with real-time/diagnostic imaging [100, 101]. Where nanoparticles have a dual function for imaging and therapy, they are increasingly known as theranostic (i.e. *thera*py + diag*nostic*) nanoparticles. Moreover, what was achieved with inorganic "hard" iron oxide nanoparticles was subsequently reported using LNPs. For instance, a multimodal imaging theranostic siRNA-**ABC** nanoparticle system was recently described that had been assembled by the stepwise formulation of PEGylated cationic liposomes (prepared using Gd.DOTA.DSA and DOPE-Rhoda amongst other lipids), followed by the encapsulation of Alexa fluor 488-labelled anti-survivin siRNA. These multimodal imaging theranostic nanoparticles were found able to mediate functional delivery of siRNA to tumours giving rise to a significant phenotypic (pharmacodynamic) reductions in tumour sizes relative to controls, while at the same time nanoparticle biodistribution (DOPE-Rhoda fluorescence plus MRI), and siRNA pharmacokinetic behaviour (Alexa fluor 488 fluorescence) could be observed by means of simultaneous real-time imaging [65]. This concept of multimodal imaging theranostic nanoparticles for cancer imaging and therapy is certain to grow in importance in preclinical cancer nanotechnology studies and maybe in the clinic too.

18.3   Nanoparticle Applications in Triggered and Image-Guided Therapies

Multimodal imaging theranostic nanoparticles may offer substantial benefits for cancer diagnosis and therapy going forward, but only in combination with further advances in nanoparticle platform delivery technologies. What might these advances be and how might they be implemented? As far as imaging nanoparticles are concerned for detection of cancer, provided that all that is required for diagnosis is nanoparticle accumulation within cancer lesions, then current imaging nanoparticle technologies may well be sufficient. However, for personalized medicine to really take off, the detection of cancer disease specific biomarkers in vivo is really required. In order to achieve this, considerable attention may well have to be paid to the appropriate design and selection of ligands (**D**-components) for the biological targeting layer (or **D**-layer).

As far as nanoparticles for cancer therapy are concerned, the opportunities for delivery are relatively limited at this point in time, primarily due to the facile partition of current nanoparticles post-administration to liver and to solid tumours in vivo and in clinic. In order to enable partition to other organs of interest and even to diseased target cell populations within, there is now an imperative to introduce new design features involving new tool-kits of chemical components. Moreover the **ABCD** nanoparticle paradigm itself has one primary design weakness in that the stealth/biocompatibility polymer layer (or **C**-layer) (typically PEG, main **C**-component) does not prevent nanoparticle entry into cells but may substantially inhibit functional intracellular delivery of the therapeutic agent, unless sufficiently removed by the time of target cell-entry or else during the process of cell-entry. Learning the rules for the control of nanoparticle biodistribution and of therapeutic agent cargo pharmacokinetics may take several years yet even though rule sets are emerging. Therefore, overcoming the **C**-layer paradox should be the primary focus for therapeutic nanoparticle development over the next few years. Accordingly, there has been a growing interest in the concept of nanoparticles that possess the property of triggerability. Such nanoparticles are designed for high levels of stability in biological fluids from points of administration to target cells whereupon they become triggered for the controlled release of entrapped therapeutic agent payload(s) by changes in local endogenous conditions (such as pH, $t_{1/2}$, enzyme, redox state, and temperature status) [61–66, 102], or through application of an external/exogenous stimulus (Rosca et al. 2014, manuscript in submission). While much of previous work on this topic has revolved around change(s) in local endogeneous conditions [61–66, 102], the development of appropriate exogenous stimuli looks to be a real growth area for the future. In principle, all **ABC** and **ABCD** nanoparticle systems could be triggered to exhibit physical property change(s) appropriate for controlled release through interaction with light, ultrasound, radiofrequency and thermal radiation from defined sources. So how might this be harnessed using "soft" organic and "hard" inorganic nanoparticles?

Today the journey towards "soft" organic LNPs for cancer therapy that can be described as truely triggered multimodal imaging theranostic drug-nanoparticles appears well underway. A few years ago thermally triggered drug-**ABC** nanoparticles (now known as Thermodox®, Celsion) were described based upon Doxil®. Thermodox® nanoparticles are formulated with lipids that included lyso-phospholipids in order to encapsulate doxorubicin within thermosensitive, nanoparticle lipid bilayer membranes [103, 104]. At induced temperatures above 37 °C, these membranes appear to become unexpectedly porous allowing for substantial local controlled release of drug. Needham et al. were first to demonstrate the use of such thermally triggered drug-**ABC** nanoparticles for the controlled local release of drug into target tissues in vivo [105], thus allowing for the potential treatment of tumours more efficiently than was achieved following administration of the thermally insensitive, Doxil® parent system [106]. Thermodox® is currently the subject of phase III HEAT studies and phase II ABLATE studies. In the latter studies, Thermodox® is administered intravenously in combination with Radio Frequency Ablation (RFA) of tumour tissue. In this case, the RFA acts as an exogenous source of local tissue hyperthermia (39.5–42 °C) that simultaneously acts as a thermal trigger for controlled release of encapsulated doxorubicin from the central aqueous cavity of Thermodox® nanoparticles. The company's pipeline going forward is focused on the use of Thermodox® nanoparticles under thermal triggered release conditions for the treatment of breast, colorectal and primary liver cancer lesions [107, 108]. This is the first time that thermally triggered drug-**ABC** nanoparticles have been devised and used in clinical trials.

A further evolution of this concept has now been more recently reported with the simultaneous entrapment of both doxorubicin and a MRI positive contrast agent, $Gd(HPDO_3A)(H_2O)$, into thermally triggered drug-**ABC** nanoparticles [109]. High Frequency Ultrasound (HIFU) was used as an alternative thermal trigger for the controlled release of encapsulated drug at 42 °C. The simultaneous release of MRI contrast agent enabled the observation of release in real time and led to an estimation of doxorubicin nanoparticle release kinetics. Researchers involved in Thermodox® have similarly reported on the development of thermally triggered drug-**ABC** nanoparticles with co-encapsulated doxorubicin and the MRI contrast agent Prohance® [110]. Using HIFU as a thermal trigger once more, they were able to promote controlled release of drug in rabbits with Vx2 tumours, and monitor drug release in real time by MRI [111]. The same researchers also developed an algorithm to simulate the thermal trigger effects of HIFU [112]. Simulation data were in agreement with mean tissue temperature increases from 37 °C to between 40.4 °C and 41.3 °C, resulting in quite heterogeneous drug release kinetic behaviour [112]. By contrast, we have striven to draw inspiration from the Gd-**ABC** and Gd-**ABCD** imaging nanoparticle systems described above [83–85, 113, 114], and Thermodox® data, in order to derive thermally triggered theranostic drug-**ABC** nanoparticles. These might also be described as thermal trig-anostic drug-**ABC** nanoparticles (shortened to the acronym thermal TNPs) (Fig. 18.3). By description, these nanoparticles are enabled for thermally triggered release of encapsulated drug in tumours by means of ultrasound together with real time, diagnostic imaging of nanoparticle

Trig-anostic drug-ABC nanoparticles; *design principles*

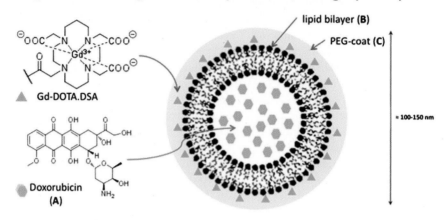

- **PEG-coat:** at least 4 mol% to give good *in vivo* stability
- **MRI-label:** Gd-DOTA.DSA to minimise Gd^{3+} leeching risk, surface attached for best contrast
- **Doxorubicin loaded:** to highest capacity possible
- **Thermal triggered release:** between 39 - 45°C, minimal release at 37°C
- **Size:** 100-150 nm to allow tumour enhanced uptake

Fig. 18.3 Schematic of thermal trig-anostic drug-**ABC** nanoparticles (thermal TNPs) enabled for thermally triggered release of encapsulated drug in tumours by means of ultrasound together with real time, diagnostic imaging of nanoparticle biodistribution by MRI with drug pharmacokinetics

biodistribution with drug pharmacokinetics. Critical to this proposition is the use of Gd.DOTA.DSA once again. Going forward, lipidic MRI agent use should be supplemented with other imaging agents leading to new families of triggered multimodal imaging theranostic drug-**ABC** nanoparticles. These could also be described as trig-anostic^n drug-**ABC** nanoparticles where n is number of imaging modes employed, a description that could also be shortened to the acronym ^{n}TNPs.

In the case of "hard" inorganic nanoparticle systems, gold nanoparticles provide for a useful illustration. These belong to a class of nanoparticles known as nanoshells with tunable optical resonances. These nanoshells consist of a core, in this case silica that is surrounded by a thin metal shell, in this case gold [115]. These particles exhibit highly tunable surface plasmon resonances that absorb NIR radiation from a bespoke laser source and then transmit locally causing local tissue damage while leaving surrounding tissue intact [116]. Nanoshells are currently under evaluation in a number of clinical settings after a 5 years period of intensive preclinical development [117]. Obviously, in this instance, nanoshells are triggered to act in effect as their own "therapeutic agent", but nanoshells can also be administered in combination with anti-cancer therapeutic antibodies opening up options of combining anti-cancer antibody therapy with hyperthermia therapy [118]. In hyperthermia treatment, nanoshells may be replaced shortly by nanorods in the next steps of development in these "hard" nanoparticle systems [119].

A peak of design must then be represented by the development of targeted trig-anostic^n therapeutically multifunctional drug-**ABCD** nanoparticles. These might also be described as targeted trig-anostic^n drug^m-**ABCD** nanoparticles, where n is number of imaging modes employed in nanoparticle and m is the number of active therapeutic agents (APIs) encapsulated/entrapped, a description that reduces to the corresponding acronym of targeted ^nT_mNPs. Amazingly, while LNP and PNP systems of this type have yet to be devised, nanoshell structures have now been reported that have been pre-doped with MRI probes (by introduction of a 10 nm iron oxide layer over the silica core) and/or NIR probes (indocyanine green dye), then set up (with streptavidin) for surface conjugation of anti-HER2 antibodies (biotin labelled) with an additional surface PEG biocompatibility layer (introduced by disulphide post coupling bond formation). Such ensembles can be described readily as targeted trig-anostic^2 drug^2-**ABCD** nanoparticle systems (or targeted $^2T_2$NPs) enabled for real time/diagnostic bimodal MRI and NIR contrast imaging accessed in combination with the capability for dual targeted and triggered chemotherapy (by anti-HER-2 antibodies) and photo-thermal ablation therapy (post illumination with a 808 nm wavelength NIR laser) either in vitro or in vivo [120, 121].

18.4 Conclusions and Future Perspective

Nanotechnology is revolutionising research and development in healthcare. Currently, the most advanced clinical-grade nanotechnologies in cancer are lipid-based and some "hard inorganic" nanoparticles. Recent studies show more evidence that biocompatibility and safety of nanoparticles depends on the material, and surface chemistry properties. Even quantum dots that have been previously characterised as toxic can be adapted for apparently safe use in non-human primates [122]. Unfortunately, there is still some scepticism from the big pharma industry and from clinicians themselves regarding the efficacy and safety of nanoparticle technologies. Such scepticism will only be solved with the advent of reliable cGMP-grade manufacturing processes and reliable preclinical ADME/toxicology data, followed by a range of successful first in man-studies. While these data are being acquired, nanoparticle technologies continue to be innovated in the laboratory. In this case, there appears to be an increasing push towards targeted trig-anostic^n drug^m-**ABCD** nanoparticles (^nT_mNPs) enabled for both targeted and triggered release of m active therapeutic agents (APIs) (including small molecule drug entities), all monitored simultaneously by real time/diagnostic imaging using n different imaging agent probes integrated into individual nanoparticles. Of the latter, both NIR and ^{19}F-NMR spectroscopy probes [123], could have real clinical potential alongside MRI. Such functional multiplicity offers the very real opportunity for highly personalized, hyper-functionalized drug-nanoparticles tailor-made (designed and assembled) from tool-kits of chemical components that have themselves been highly refined for specific, personalized delivery applications. As this vision takes shape, so we will be looking on a very different world of innovative, interactive healthcare products with vastly more potential to treat and even to cure cancer than has ever been seen before.

And what of routine personalized cancer diagnosis and therapy? Do current advances in nanoparticle development allow us to close the virtuous circle of molecular profiling to personalized cancer nanomedicine? At this stage the answer must be, "not yet" or "status unproven". Clearly cancer imaging and therapy using nanoparticle technologies looks and is entirely becoming clinically realistic. But we are not yet at the point where patient specific, cancer disease-specific biomarkers can be detected in vivo using nanotechnology followed in the clinic by nanoparticle-mediated functional delivery of biomarker specific therapeutic agents. However, at least where ncRNAs are concerned, the prospect of such a cycle does appear imminent. As ncRNA profiling of cancers take place, so one can envisage a time when the follow on design of nanoparticles for the functional delivery of RNAi effectors targeted against specific cancer biomarker ncRNAs could become routine. Once this can be achieved, then the virtuous circle of personalized cancer nanomedicine will be properly closed.

References

1. Ferrari M (2005) Cancer nanotechnology: opportunities and challenges. Nat Rev Cancer 5(3):161–171. doi:10.1038/nrc1566
2. Srinivas PR, Barker P, Srivastava S (2002) Nanotechnology in early detection of cancer. Lab Investig J Tech Methods Pathol 82(5):657–662
3. Nie S, Xing Y, Kim GJ, Simons JW (2007) Nanotechnology applications in cancer. Annu Rev Biomed Eng 9:257–288. doi:10.1146/annurev.bioeng.9.060906.152025
4. Wang MD, Shin DM, Simons JW, Nie S (2007) Nanotechnology for targeted cancer therapy. Expert Rev Anticancer Ther 7(6):833–837. doi:10.1586/14737140.7.6.833
5. Anon (2007) Cancer nanotechnology: small, but heading for the big time. Nat Rev Drug Discov 6(3):174–175. doi:10.1038/nrd2285
6. EPT Nanomedicine (2009) Roadmaps in nanomedicine towards 2020. http://www.etp-nanomedicine.eu/public/press-documents/publications/etpn-publications/091022_ETPN_Report_2009.pdf. Accessed 22 June 2014
7. Golub TR, Slonim DK, Tamayo P, Huard C, Gaasenbeek M, Mesirov JP, Coller H, Loh ML, Downing JR, Caligiuri MA, Bloomfield CD, Lander ES (1999) Molecular classification of cancer: class discovery and class prediction by gene expression monitoring. Science 286(5439):531–537
8. Ross DT, Scherf U, Eisen MB, Perou CM, Rees C, Spellman P, Iyer V, Jeffrey SS, Van de Rijn M, Waltham M, Pergamenschikov A, Lee JC, Lashkari D, Shalon D, Myers TG, Weinstein JN, Botstein D, Brown PO (2000) Systematic variation in gene expression patterns in human cancer cell lines. Nat Genet 24(3):227–235. doi:10.1038/73432
9. Scherf U, Ross DT, Waltham M, Smith LH, Lee JK, Tanabe L, Kohn KW, Reinhold WC, Myers TG, Andrews DT, Scudiero DA, Eisen MB, Sausville EA, Pommier Y, Botstein D, Brown PO, Weinstein JN (2000) A gene expression database for the molecular pharmacology of cancer. Nat Genet 24(3):236–244. doi:10.1038/73439
10. Rubin MA (2001) Use of laser capture microdissection, cDNA microarrays, and tissue microarrays in advancing our understanding of prostate cancer. J Pathol 195(1):80–86. doi:10.1002/path.892

11. Evans WE, Relling MV (2004) Moving towards individualized medicine with pharmacogenomics. Nature 429(6990):464–468. doi:10.1038/nature02626

12. Petros WP, Evans WE (2004) Pharmacogenomics in cancer therapy: is host genome variability important? Trends Pharmacol Sci 25(9):457–464. doi:10.1016/j.tips.2004.07.007

13. Hwang SR, Ku SH, Joo MK, Kim SH, Kwon IC (2014) Theranostic nanomaterials for image-guided gene therapy. MRS Bull 39(01):44–50

14. Sun NF, Liu ZA, Huang WB, Tian AL, Hu SY (2014) The research of nanoparticles as gene vector for tumor gene therapy. Crit Rev Oncol Hematol 89(3):352–357

15. Yin H, Kanasty RL, Eltoukhy AA, Vegas AJ, Dorkin JR, Anderson DG (2014) Non-viral vectors for gene-based therapy. Nat Rev Genet 15(8):541–555

16. Miller AD (2008) Towards safe nanoparticle technologies for nucleic acid therapeutics. Tumori 94(2):234–245

17. Kostarelos K, Miller AD (2005) Synthetic, self-assembly ABCD nanoparticles; a structural paradigm for viable synthetic non-viral vectors. Chem Soc Rev 34(11):970–994

18. Fletcher S, Ahmad A, Perouzel E, Heron A, Miller AD, Jorgensen MR (2006) In vivo studies of dialkynoyl analogues of DOTAP demonstrate improved gene transfer efficiency of cationic liposomes in mouse lung. J Med Chem 49:349–357

19. Fletcher S, Ahmad A, Perouzel E, Jorgensen MR, Miller AD (2006) A dialkanoyl analogue of DOPE improves gene transfer of lower-charged, cationic lipoplexes. Org Biomol Chem 4:196–199

20. Fletcher S, Ahmad A, Price WS, Jorgensen MR, Miller AD (2008) Biophysical properties of CDAN/DOPE-analogue lipoplexes account for enhanced gene delivery. Chembiochem 9(3):455–463

21. Miller AD (2003) The problem with cationic liposome/micelle-based non-viral vector systems for gene therapy. Curr Med Chem 10(14):1195–1211

22. Miller AD (2004) Gene therapy needs robust synthetic nonviral platform technologies. Chembiochem 5(1):53–54

23. Miller AD (2004) Nonviral liposomes. In: Springer CJ (ed) Methods in molecular medicine, vol 90. Humana Press, Totowa, pp 107–137

24. Oliver M, Jorgensen MR, Miller AD (2004) The facile solid-phase synthesis of cholesterol-based polyamine lipids. Tetrahedron Lett 45:3105–3108

25. Spagnou S, Miller AD, Keller M (2004) Lipidic carriers of siRNA: differences in the formulation, cellular uptake, and delivery with plasmid DNA. Biochemistry 43(42): 13348–13356

26. Tagawa T, Manvell M, Brown N, Keller M, Perouzel E, Murray KD, Harbottle RP, Tecle M, Booy F, Brahimi-Horn MC, Coutelle C, Lemoine NR, Alton EWFW, Miller AD (2002) Characterisation of LMD virus-like nanoparticles self-assembled from cationic liposomes, adenovirus core peptide μ (mu) and plasmid DNA. Gene Ther 9(9):564–576

27. Bharali DJ, Khalil M, Gurbuz M, Simone TM, Mousa SA (2009) Nanoparticles and cancer therapy: a concise review with emphasis on dendrimers. Int J Nanomedicine 4:1–7

28. Cho K, Wang X, Nie S, Chen ZG, Shin DM (2008) Therapeutic nanoparticles for drug delivery in cancer. Clin Cancer Res 14(5):1310–1316. doi:10.1158/1078-0432.CCR-07-1441

29. Davis ME, Chen ZG, Shin DM (2008) Nanoparticle therapeutics: an emerging treatment modality for cancer. Nat Rev Drug Discov 7(9):771–782. doi:10.1038/nrd2614

30. Farokhzad OC, Langer R (2009) Impact of nanotechnology on drug delivery. ACS Nano 3(1):16–20. doi:10.1021/nn900002m

31. Lammers T, Hennink WE, Storm G (2008) Tumour-targeted nanomedicines: principles and practice. Br J Cancer 99(3):392–397. doi:10.1038/sj.bjc.6604483

32. Peer D, Karp JM, Hong S, Farokhzad OC, Margalit R, Langer R (2007) Nanocarriers as an emerging platform for cancer therapy. Nat Nanotechnol 2(12):751–760. doi:10.1038/nnano.2007.387

33. Tanaka T, Decuzzi P, Cristofanilli M, Sakamoto JH, Tasciotti E, Robertson FM, Ferrari M (2009) Nanotechnology for breast cancer therapy. Biomed Microdevices 11(1):49–63. doi:10.1007/s10544-008-9209-0

34. Youan BB (2008) Impact of nanoscience and nanotechnology on controlled drug delivery. Nanomedicine (Lond) 3(4):401–406. doi:10.2217/17435889.3.4.401

35. Byrne JD, Betancourt T, Brannon-Peppas L (2008) Active targeting schemes for nanoparticle systems in cancer therapeutics. Adv Drug Deliv Rev 60(15):1615–1626. doi:10.1016/j. addr.2008.08.005

36. Zolnik BS, Sadrieh N (2009) Regulatory perspective on the importance of ADME assessment of nanoscale material containing drugs. Adv Drug Deliv Rev 61(6):422–427. doi:10.1016/j. addr.2009.03.006

37. Matsumura Y, Maeda H (1986) A new concept for macromolecular therapeutics in cancer chemotherapy: mechanism of tumoritropic accumulation of proteins and the antitumor agent smancs. Cancer Res 46(12 Pt 1):6387–6392

38. Kobayashi H, Watanabe R, Choyke PL (2014) Improving conventional enhanced permeability and retention (EPR) effects; what is the appropriate target? Theranostics 4(1):81–89

39. Thanou M, Duncan R (2003) Polymer-protein and polymer-drug conjugates in cancer therapy. Curr Opin Investig Drugs 4(6):701–709

40. Cohen BE, Bangham AD (1972) Diffusion of small non-electrolytes across liposome membranes. Nature 236(5343):173–174

41. Johnson SM, Bangham AD (1969) Potassium permeability of single compartment liposomes with and without valinomycin. Biochim Biophys Acta 193(1):82–91

42. Lasic D (ed) (1998) Medical applications of liposomes. Elsevier, Amsterdaam

43. Lin Q, Chen J, Zhang Z, Zheng G (2014) Lipid-based nanoparticles in the systemic delivery of siRNA. Nanomedicine 9(1):105–120

44. Allen TM, Cullis PR (2004) Drug delivery systems: entering the mainstream. Science 303(5665):1818–1822. doi:10.1126/science.1095833

45. Torchilin VP (2005) Recent advances with liposomes as pharmaceutical carriers. Nat Rev Drug Discov 4(2):145–160. doi:10.1038/nrd1632

46. Forssen EA, Tokes ZA (1983) Improved therapeutic benefits of doxorubicin by entrapment in anionic liposomes. Cancer Res 43(2):546–550

47. Treat J, Greenspan A, Forst D, Sanchez JA, Ferrans VJ, Potkul LA, Woolley PV, Rahman A (1990) Antitumor activity of liposome-encapsulated doxorubicin in advanced breast cancer: phase II study. J Natl Cancer Inst 82(21):1706–1710

48. Robert NJ, Vogel CL, Henderson IC, Sparano JA, Moore MR, Silverman P, Overmoyer BA, Shapiro CL, Park JW, Colbern GT, Winer EP, Gabizon AA (2004) The role of the liposomal anthracyclines and other systemic therapies in the management of advanced breast cancer. Semin Oncol 31(6 Suppl 13):106–146

49. Straubinger RM, Lopez NG, Debs RJ, Hong K, Papahadjopoulos D (1988) Liposome-based therapy of human ovarian cancer: parameters determining potency of negatively charged and antibody-targeted liposomes. Cancer Res 48(18):5237–5245

50. Allen TM, Martin FJ (2004) Advantages of liposomal delivery systems for anthracyclines. Semin Oncol 31(6 Suppl 13):5–15

51. Gabizon A, Martin F (1997) Polyethylene glycol-coated (pegylated) liposomal doxorubicin. Rationale for use in solid tumours. Drugs 54(Suppl 4):15–21

52. Gabizon A, Shmeeda H, Barenholz Y (2003) Pharmacokinetics of pegylated liposomal Doxorubicin: review of animal and human studies. Clin Pharmacokinet 42(5):419–436

53. Papahadjopoulos D, Allen TM, Gabizon A, Mayhew E, Matthay K, Huang SK, Lee KD, Woodle MC, Lasic DD, Redemann C et al (1991) Sterically stabilized liposomes: improvements in pharmacokinetics and antitumor therapeutic efficacy. Proc Natl Acad Sci U S A 88(24):11460–11464

54. Woodle MC, Lasic DD (1992) Sterically stabilized liposomes. Biochim Biophys Acta 1113(2):171–199

55. Damascelli B, Cantu G, Mattavelli F, Tamplenizza P, Bidoli P, Leo E, Dosio F, Cerrotta AM, Di Tolla G, Frigerio LF, Garbagnati F, Lanocita R, Marchiano A, Patelli G, Spreafico C, Ticha V, Vespro V, Zunino F (2001) Intraarterial chemotherapy with polyoxyethylated castor oil free paclitaxel, incorporated in albumin nanoparticles (ABI-007): phase II study of patients

with squamous cell carcinoma of the head and neck and anal canal: preliminary evidence of clinical activity. Cancer 92(10):2592–2602

56. Desai N, Trieu V, Yao Z, Louie L, Ci S, Yang A, Tao C, De T, Beals B, Dykes D, Noker P, Yao R, Labao E, Hawkins M, Soon-Shiong P (2006) Increased antitumor activity, intratumor paclitaxel concentrations, and endothelial cell transport of cremophor-free, albumin-bound paclitaxel, ABI-007, compared with cremophor-based paclitaxel. Clin Cancer Res 12(4):1317–1324. doi:10.1158/1078-0432.CCR-05-1634

57. Miele E, Spinelli GP, Miele E, Tomao F, Tomao S (2009) Albumin-bound formulation of paclitaxel (Abraxane ABI-007) in the treatment of breast cancer. Int J Nanomedicine 4:99–105

58. Vogel SM, Minshall RD, Pilipovic M, Tiruppathi C, Malik AB (2001) Albumin uptake and transcytosis in endothelial cells in vivo induced by albumin-binding protein. Am J Physiol Lung Cell Mol Physiol 281(6):L1512–L1522

59. Hawkins MJ, Soon-Shiong P, Desai N (2008) Protein nanoparticles as drug carriers in clinical medicine. Adv Drug Deliv Rev 60(8):876–885. doi:10.1016/j.addr.2007.08.044

60. Davis ME (2009) The first targeted delivery of siRNA in humans via a self-assembling, cyclodextrin polymer-based nanoparticle: from concept to clinic. Mol Pharm 6(3):659–668. doi:10.1021/mp900015y

61. Aissaoui A, Chami M, Hussein M, Miller AD (2011) Efficient topical delivery of plasmid DNA to lung in vivo mediated by putative triggered, PEGylated pDNA nanoparticles. J Control Release 154:275–284

62. Carmona S, Jorgensen MR, Kolli S, Crowther C, Salazar FH, Marion PL, Fujino M, Natori Y, Thanou M, Arbuthnot P, Miller AD (2009) Controlling HBV replication in vivo by intravenous administration of triggered PEGylated siRNA-nanoparticles. Mol Pharm 6(3):706–717

63. Drake CR, Aissaoui A, Argyros O, Serginson JM, Monnery BD, Thanou M, Steinke JHG, Miller AD (2010) Bioresponsive small molecule polyamines as non-cytotoxic alternative to polyethylenimine. Mol Pharm 7(6):2040–2055

64. Drake CR, Aissaoui A, Argyros O, Thanou M, Steinke JH, Miller AD (2013) Examination of the effect of increasing the number of intra-disulfide amino functional groups on the performance of small molecule cyclic polyamine disulfide vectors. J Control Release 171(1):81–90. doi:10.1016/j.jconrel.2013.02.014

65. Kenny GD, Kamaly N, Kalber TL, Brody LP, Sahuri M, Shamsaei E, Miller AD, Bell JD (2011) Novel multifunctional nanoparticle mediates siRNA tumour delivery, visualisation and therapeutic tumour reduction in vivo. J Control Release 149(2):111–116. doi:10.1016/j.jconrel.2010.09.020

66. Kolli S, Wong SP, Harbottle R, Johnston B, Thanou M, Miller AD (2013) pH-triggered nanoparticle mediated delivery of siRNA to liver cells in vitro and in vivo. Bioconjug Chem 24(3):314–332. doi:10.1021/bc3004099

67. Mével M, Kamaly N, Carmona S, Oliver MH, Jorgensen MR, Crowther C, Salazar FH, Marion PL, Fujino M, Natori Y, Thanou M, Arbuthnot P, Yaouanc J-J, Jaffres PA, Miller AD (2010) DODAG; a versatile new cationic lipid that mediates efficient delivery of pDNA and siRNA. J Control Release 143:222–232

68. Andreu A, Fairweather N, Miller AD (2008) Clostridium neurotoxin fragments as potential targeting moieties for liposomal gene delivery to the CNS. Chembiochem 9(2):219–231

69. Chen J, Jorgensen MR, Thanou M, Miller AD (2011) Post-coupling strategy enables true receptor-targeted nanoparticles. J RNAi Gene Silenc Int j RNA Gene Target Res 7:449–455

70. Wang M, Lowik DW, Miller AD, Thanou M (2009) Targeting the urokinase plasminogen activator receptor with synthetic self-assembly nanoparticles. Bioconjug Chem 20(1):32–40

71. Wang M, Miller AD, Thanou M (2013) Effect of surface charge and ligand organization on the specific cell-uptake of uPAR-targeted nanoparticles. J Drug Target 21(7):684–692. doi:10.3109/1061186X.2013.805336

72. Drummond DC, Noble CO, Guo Z, Hong K, Park JW, Kirpotin DB (2006) Development of a highly active nanoliposomal irinotecan using a novel intraliposomal stabilization strategy. Cancer Res 66(6):3271–3277. doi:10.1158/0008-5472.CAN-05-4007

73. Flexman JA, Yung A, Yapp DT, Ng SS, Kozlowski P (2008) Assessment of vessel size by MRI in an orthotopic model of human pancreatic cancer. Conf Proc IEEE Eng Med Biol Soc 2008:851–854. doi:10.1109/IEMBS.2008.4649287

74. Ting G, Chang CH, Wang HE (2009) Cancer nanotargeted radiopharmaceuticals for tumor imaging and therapy. Anticancer Res 29(10):4107–4118

75. Harrington KJ, Mohammadtaghi S, Uster PS, Glass D, Peters AM, Vile RG, Stewart JS (2001) Effective targeting of solid tumors in patients with locally advanced cancers by radiolabeled pegylated liposomes. Clin Cancer Res 7(2):243–254

76. Erdogan S (2009) Liposomal nanocarriers for tumor imaging. J Biomed Nanotechnol 5(2): 141–150

77. Mulder WJ, Strijkers GJ, van Tilborg GA, Griffioen AW, Nicolay K (2006) Lipid-based nanoparticles for contrast-enhanced MRI and molecular imaging. NMR Biomed 19(1): 142–164. doi:10.1002/nbm.1011

78. Devoisselle JM, Vion-Dury J, Galons JP, Confort-Gouny S, Coustaut D, Canioni P, Cozzone PJ (1988) Entrapment of gadolinium-DTPA in liposomes. Characterization of vesicles by P-31 NMR spectroscopy. Invest Radiol 23(10):719–724

79. Unger E, Fritz T, Shen DK, Wu G (1993) Manganese-based liposomes. Comparative approaches. Invest Radiol 28(10):933–938

80. Kozlowska D, Foran P, MacMahon P, Shelly MJ, Eustace S, O'Kennedy R (2009) Molecular and magnetic resonance imaging: the value of immunoliposomes. Adv Drug Deliv Rev 61(15):1402–1411. doi:10.1016/j.addr.2009.09.003

81. Kabalka GW, Davis MA, Buonocore E, Hubner K, Holmberg E, Huang L (1990) Gd-labeled liposomes containing amphipathic agents for magnetic resonance imaging. Invest Radiol 25(Suppl 1):S63–S64

82. van Tilborg GA, Strijkers GJ, Pouget EM, Reutelingsperger CP, Sommerdijk NA, Nicolay K, Mulder WJ (2008) Kinetics of avidin-induced clearance of biotinylated bimodal liposomes for improved MR molecular imaging. Magn Reson Med 60(6):1444–1456. doi:10.1002/mrm.21780

83. Kamaly N, Kalber T, Kenny G, Bell J, Jorgensen M, Miller A (2010) A novel bimodal lipidic contrast agent for cellular labelling and tumour MRI. Org Biomol Chem 8(1):201–211. doi:10.1039/b910561a

84. Kamaly N, Kalber T, Thanou M, Bell JD, Miller AD (2009) Folate receptor targeted bimodal liposomes for tumor magnetic resonance imaging. Bioconjug Chem 20(4):648–655. doi:10.1021/bc8002259

85. Kamaly N, Kalber T, Ahmad A, Oliver MH, So PW, Herlihy AH, Bell JD, Jorgensen MR, Miller AD (2008) Bimodal paramagnetic and fluorescent liposomes for cellular and tumor magnetic resonance imaging. Bioconjug Chem 19(1):118–129. doi:10.1021/bc7001715

86. Almeida AJ, Souto E (2007) Solid lipid nanoparticles as a drug delivery system for peptides and proteins. Adv Drug Deliv Rev 59(6):478–490. doi:10.1016/j.addr.2007.04.007

87. Cavalli R, Caputo O, Carlotti ME, Trotta M, Scarnecchia C, Gasco MR (1997) Sterilization and freeze-drying of drug-free and drug-loaded solid lipid nanoparticles. Int J Pharm 148(1):47–54

88. Muller RH, Mehnert W, Lucks JS (1995) Solid lipid nanoparticles (Sln) – an alternative colloidal carrier system for controlled drug-delivery. Eur J Pharm Biopharm 41(1):62–69

89. Morel S, Terreno E, Ugazio E, Aime S, Gasco MR (1998) NMR relaxometric investigations of solid lipid nanoparticles (SLN) containing gadolinium(III) complexes. Eur J Pharm Biopharm 45(2):157–163

90. Tomalia DA, Uppuluri S, Swanson DR (1999) Dendritic macromolecules: a fourth major class of polymer architecture – new properties driven by architecture. Mater Res Soc Symp Proc 543:289–298

91. Hawker CJ, Frechet JMJ (1990) Preparation of polymers with controlled molecular architecture – a new convergent approach to dendritic macromolecules. J Am Chem Soc 112(21):7638–7647

92. Svenson S, Tomalia DA (2005) Dendrimers in biomedical applications–reflections on the field. Adv Drug Deliv Rev 57(15):2106–2129. doi:10.1016/j.addr.2005.09.018

93. Wolinsky JB, Grinstaff MW (2008) Therapeutic and diagnostic applications of dendrimers for cancer treatment. Adv Drug Deliv Rev 60(9):1037–1055. doi:10.1016/j.addr.2008.02.012

94. Barth RF, Coderre JA, Vicente MG, Blue TE (2005) Boron neutron capture therapy of cancer: current status and future prospects. Clin Cancer Res 11(11):3987–4002. doi:10.1158/1078-0432.CCR-05-0035

95. Wiener EC, Brechbiel MW, Brothers H, Magin RL, Gansow OA, Tomalia DA, Lauterbur PC (1994) Dendrimer-based metal chelates: a new class of magnetic resonance imaging contrast agents. Magn Reson Med 31(1):1–8

96. Harisinghani MG, Saksena MA, Hahn PF, King B, Kim J, Torabi MT, Weissleder R (2006) Ferumoxtran-10-enhanced MR lymphangiography: does contrast-enhanced imaging alone suffice for accurate lymph node characterization? AJR Am J Roentgenol 186(1):144–148. doi:10.2214/AJR.04.1287

97. Sharma R, Saini S, Ros PR, Hahn PF, Small WC, de Lange EE, Stillman AE, Edelman RR, Runge VM, Outwater EK, Morris M, Lucas M (1999) Safety profile of ultrasmall superparamagnetic iron oxide ferumoxtran-10: phase II clinical trial data. J Magn Reson Imagin JMRI 9(2):291–294

98. Islam T, Harisinghani MG (2009) Overview of nanoparticle use in cancer imaging. Cancer Biomark Sect A Dis Markers 5(2):61–67. doi:10.3233/CBM-2009-0578

99. Yu MK, Jeong YY, Park J, Park S, Kim JW, Min JJ, Kim K, Jon S (2008) Drug-loaded superparamagnetic iron oxide nanoparticles for combined cancer imaging and therapy in vivo. Angew Chem Int Ed Engl 47(29):5362–5365. doi:10.1002/anie.200800857

100. Medarova Z, Pham W, Farrar C, Petkova V, Moore A (2007) In vivo imaging of siRNA delivery and silencing in tumors. Nat Med 13(3):372–377. doi:10.1038/nm1486

101. Moore A, Medarova Z (2009) Imaging of siRNA delivery and silencing. Methods Mol Biol 487:93–110. doi:10.1007/978-1-60327-547-7_5

102. Yingyuad P, Mevel M, Prata C, Furegati S, Kontogiorgis C, Thanou M, Miller AD (2013) Enzyme-triggered PEGylated pDNA-nanoparticles for controlled release of pDNA in tumors. Bioconjug Chem 24(3):343–362. doi:10.1021/bc300419g

103. Kong G, Anyarambhatla G, Petros WP, Braun RD, Colvin OM, Needham D, Dewhirst MW (2000) Efficacy of liposomes and hyperthermia in a human tumor xenograft model: importance of triggered drug release. Cancer Res 60(24):6950–6957

104. Poon RT, Borys N (2009) Lyso-thermosensitive liposomal doxorubicin: a novel approach to enhance efficacy of thermal ablation of liver cancer. Expert Opin Pharmacother 10(2):333–343. doi:10.1517/14656560802677874

105. Needham D, Anyarambhatla G, Kong G, Dewhirst MW (2000) A new temperature-sensitive liposome for use with mild hyperthermia: characterization and testing in a human tumor xenograft model. Cancer Res 60(5):1197–1201

106. Needham D, Dewhirst MW (2001) The development and testing of a new temperature-sensitive drug delivery system for the treatment of solid tumors. Adv Drug Deliv Rev 53(3):285–305

107. Thomas MB, Jaffe D, Choti MM, Belghiti J, Curley S, Fong Y, Gores G, Kerlan R, Merle P, O'Neil B, Poon R, Schwartz L, Tepper J, Yao F, Haller D, Mooney M, Venook A (2010) Hepatocellular carcinoma: consensus recommendations of the National Cancer Institute Clinical Trials Planning Meeting. J Clin Oncol 28(25):3994–4005. doi:10.1200/JCO.2010.28.7805

108. Wood BJ, Poon RT, Locklin JK, Dreher MR, Ng KK, Eugeni M, Seidel G, Dromi S, Neeman Z, Kolf M, Black CD, Prabhakar R, Libutti SK (2012) Phase I study of heat-deployed liposomal doxorubicin during radiofrequency ablation for hepatic malignancies. J Vasc Interv Radiol 23(2):248–255 e247. S1051-0443(11)01427-8 [pii]. doi:10.1016/j.jvir.2011.10.018

109. de Smet M, Langereis S, van den Bosch S, Grull H (2010) Temperature-sensitive liposomes for doxorubicin delivery under MRI guidance. J Control Release 143(1):120–127. doi:10.1016/j.jconrel.2009.12.002

110. Negussie AH, Yarmolenko PS, Partanen A, Ranjan A, Jacobs G, Woods D, Bryant H, Thomasson D, Dewhirst MW, Wood BJ, Dreher MR (2011) Formulation and characterisation of magnetic resonance imageable thermally sensitive liposomes for use with magnetic resonance-guided high intensity focused ultrasound. Int J Hyperthermia 27(2):140–155. doi: 10.3109/02656736.2010.528140

111. Ranjan A, Jacobs GC, Woods DL, Negussie AH, Partanen A, Yarmolenko PS, Gacchina CE, Sharma KV, Frenkel V, Wood BJ, Dreher MR (2012) Image-guided drug delivery with magnetic resonance guided high intensity focused ultrasound and temperature sensitive liposomes in a rabbit Vx2 tumor model. J Control Release 158(3):487–494. doi:10.1016/j. jconrel.2011.12.011

112. Partanen A, Yarmolenko PS, Viitala A, Appanaboyina S, Haemmerich D, Ranjan A, Jacobs G, Woods D, Enholm J, Wood BJ, Dreher MR (2012) Mild hyperthermia with magnetic resonance-guided high-intensity focused ultrasound for applications in drug delivery. Int J Hyperthermia 28(4):320–336. doi:10.3109/02656736.2012.680173

113. Kamaly N, Miller AD (2010) Paramagnetic liposome nanoparticles for cellular and tumour imaging. Int J Mol Sci 11(4):1759–1776. doi:10.3390/ijms11041759

114. Kamaly N, Miller AD, Bell JD (2010) Chemistry of tumour targeted T1 based MRI contrast agents. Curr Top Med Chem 10(12):1158–1183, BSP/ CTMC /E-Pub/-0067-10-11 [pii]

115. Averitt RD, Westcott SL, Halas NJ (1999) Linear optical properties of gold nanoshells. J Opt Soc Am B 16(10):1824–1832

116. Hirsch LR, Stafford RJ, Bankson JA, Sershen SR, Rivera B, Price RE, Hazle JD, Halas NJ, West JL (2003) Nanoshell-mediated near-infrared thermal therapy of tumors under magnetic resonance guidance. Proc Natl Acad Sci U S A 100(23):13549–13554. doi:10.1073/ pnas.2232479100

117. Lassiter JB, Aizpurua J, Hernandez LI, Brandl DW, Romero I, Lal S, Hafner JH, Nordlander P, Halas NJ (2008) Close encounters between two nanoshells. Nano Lett 8(4):1212–1218. doi:10.1021/nl080271o

118. Leung K (2004) Iron oxide-ferritin nanocages. doi:NBK61993 [bookaccession]

119. von Maltzahn G, Park JH, Agrawal A, Bandaru NK, Das SK, Sailor MJ, Bhatia SN (2009) Computationally guided photothermal tumor therapy using long-circulating gold nanorod antennas. Cancer Res 69(9):3892–3900. doi:10.1158/0008-5472.CAN-08-4242

120. Bardhan R, Chen W, Bartels M, Perez-Torres C, Botero MF, McAninch RW, Contreras A, Schiff R, Pautler RG, Halas NJ, Joshi A (2010) Tracking of multimodal therapeutic nanocomplexes targeting breast cancer in vivo. Nano Lett. doi:10.1021/nl102889y

121. Bardhan R, Lal S, Joshi A, Halas NJ (2011) Theranostic nanoshells: from probe design to imaging and treatment of cancer. Acc Chem Res 44(10):936–946. doi:10.1021/ar200023x

122. Ye L, Yong KT, Liu L, Roy I, Hu R, Zhu J, Cai H, Law WC, Liu J, Wang K, Liu J, Liu Y, Hu Y, Zhang X, Swihart MT, Prasad PN (2012) A pilot study in non-human primates shows no adverse response to intravenous injection of quantum dots. Nat Nanotechnol 7(7):453–458. doi:10.1038/nnano.2012.74

123. Chen J, Lanza GM, Wickline SA (2010) Quantitative magnetic resonance fluorine imaging: today and tomorrow. Wiley Interdiscip Rev Nanomed Nanobiotechnol 2(4):431–440. doi:10.1002/wnan.87

Chapter 19
Challenges to Nanomedicine

Richard Moore

19.1 Introduction

The recent past has seen a period of considerable financial upheaval and constraint that has affected healthcare and healthcare provision like many other sectors. There is an increasing pressure on those bringing forward new medical technologies to ensure that they are capable of outperforming existing, established technologies, that they have a high benefit-to-risk ratio and that they are affordable or can otherwise lead to cost-savings in healthcare systems where resource availability is a constant concern.

While economic factors are particularly sensitive in the current financial climate, there are a number of other important hurdles to be negotiated in bringing any new medical technology to the clinic. These include

- taking account of demographic trends and associated changes in healthcare priorities
- addressing and minimising risks
- understanding which regulatory system(s) apply and ensuring product compliance
- understanding and negotiating reimbursement systems
- preparing for healthcare technology assessment
- considering the impact that emerging technologies may have on established medical practice
- ensuring that there is professional uptake of new technologies and addressing training issues that may arise
- addressing public understanding and perception issues
- in some cases, addressing new ethical challenges that the technology may bring

R. Moore (✉)
Biomimesis, Melton Mowbray, LE13 OUH, UK
e-mail: richard@biomimesis.net

© Springer Science+Business Media New York 2014
Y. Ge et al. (eds.), *Nanomedicine*, Nanostructure Science and Technology,
DOI 10.1007/978-1-4614-2140-5_19

While some of these topics, such as risk and ethical issues, are explored in greater depth in other chapters, they are reviewed and discussed in the following sections in order to provide a broad overview of some of the important challenges and milestones towards successful commercialization and utilization of medical products based on nanotechnologies.

19.2   The Rising Costs of Healthcare

It is estimated that, in all countries, both health and long-term care will drive up public spending. In the recent OECD Economic Report No. 6, De la Maisonneuve and Oliveira Martins (2013) project that, for OECD countries, average public healthcare expenditure will increase from 5.5 % of GDP in 2010 to 8 % of GDP in 2060; whereas public long-term care expenditure will increase from 0.8 % to 1.6 % of GDP in 2060 [1]. The report projects that healthcare spending will be pushed up mostly by the combined effect of technology, relative prices, and factors such as institutions and policies, while pressures on long-term care costs will originate mostly from weaker productivity gains than in the economy as a whole.

Given the competing pressures from other social spending programmes, the report concludes that projected trends in public health and long-term care spending are likely to be a major source of concern for most governments.

A key challenge for nanomedicine will be to demonstrate that it can contribute towards containing these rising costs. Given that the implementation of technology is frequently cited as contributing to rising healthcare costs this may at first seem paradoxical. However, a major component of the cost of healthcare is hospitalization and reducing the length of stay (LoS) in hospital is a major objective for new treatments and for healthcare planners. Nanomedicine may be able to contribute to reducing the duration of in-patient stays, or to eliminating them altogether, in various ways including:

- facilitating earlier, faster or more accurate diagnosis thereby potentially reducing the length of treatment required;
- contributing to the efficacy of treatment and improving the prognosis for the patient;
- facilitating treatment at home, at the GP's surgery or as an outpatient;
- improving the performance of individual drugs and medical devices;
- contributing to personalized medicine, e.g. by selecting and tailoring treatment to suit the individual patient and their condition.

Robinson and Smith [2] suggest, for example, that there are numerous examples of new products and processes in healthcare that reduce rather than increase the rate of spending growth and that, without these, total costs would be increasing even more rapidly than they are. These include:

- innovative new drugs, tests, devices, and other products (as distinct from services) that are cheaper to manufacture or use than those they replace;

– changes in processes that allow less highly trained but sufficiently competent workers to substitute for more highly trained and expensive staff thereby releasing them for more complex and demanding procedures. Examples could include substituting physician generalists for specialists, nurse practitioners and pharmacists for physicians, non-licensed staff for nurses, and family members and patients themselves for paid staff of any kind.

– sites of care that are less elaborate but which are adequate for the tasks under consideration, including the home itself as an effective site for care in the area of chronic illness.

They further suggest that synergies between changes in one dimension of care and changes in the others may be the most disruptive in terms of channelling patients in new directions and forcing major but desirable changes on both manufacturers and medical practitioners – more so than individual changes in products, personnel, or facilities.

Citing experience in other sectors they suggest that cost-reducing innovations are attributable to both new technology, and to new types of business model that are simpler and cheaper than those they replace, resulting in an expansion of the market due to the increased affordability of these services.

Krishna Kumar (2011), with reference to new medical technologies, makes the point that much of the effort of companies relates to providing additional features to score over their competitors' products but that matter very little in day-to-day decisions, while there is very little focus on making technology widely accessible and inexpensive [3].

Health technology assessment and reimbursement schemes, which are discussed later in this chapter, also increasingly focus on cost-containment and value-for-money. Therefore it is important that nanomedicine is able to demonstrate a contribution towards cost-containment within healthcare systems through diagnosing disease at an earlier and more treatable stage, providing more effective treatments, reducing the costs of or extending the life of products, facilitating efficiencies in the delivery of healthcare and the use of professional resources, shortening hospital stays or improving recovery times, or enabling treatment or care to be carried out in less expensive settings. In some cases, cost savings may be realized in the longer term or in parts of the healthcare system other than that where the technology is deployed, necessitating the development of a strong evidence-based case that explains the overall benefits and savings to the system.

A further factor that further exacerbates cost considerations is the demographic shift to an ageing population coupled with a reduction in the proportion of those actively contributing financially to healthcare systems.

19.3   The Demographic Shift Towards an Ageing Population

The European Commission's 2009 Ageing Report [4] estimates that, between now and 2060 within the European Union, the population will shift from a ratio of four people aged between 15 and 64 for each person aged over 65, to a ratio of only two to one.

The largest change is expected to occur between 2015 and 2035 when the current baby-boomer generation will be entering retirement. Between 2010 and 2030, the number of Europeans aged over 65 is expected to rise by nearly 40 % and, by the mid-2030s, the number of people aged 85 and over is projected to double in most European countries. Furthermore, it is estimated that around 50 % of babies born today are likely to live to 100 due to improvements in healthcare and living standards.

These demographic changes are likely to have a dramatic effect on society and to lead to new clinical challenges in relation to a wide range of health conditions associated with the elderly such as cardiovascular diseases, cancers, arthritis, osteoporosis and other orthopaedic conditions, dementias and other neurodegenerative diseases, hearing and balance disorders, and some forms of blindness.

According to the 2012 World Alzheimer Report [5] the costs associated with dementia alone were estimated to be around 1 % of the world's gross domestic product at around $604bn (€421bn) and it is likely that these costs will increase in proportion to the number of people with dementia. The report goes on to suggest that dementia, which comprises a range of neurodegenerative disorders of which Alzheimer's Disease accounts around two-thirds, poses the most significant health and social crisis of the century as its global financial burden continues to escalate, with the number of people with dementia expected to double by 2030, and more than triple by 2050. Around 682 million people will live with dementia in the next 40 years, significantly more than the population of the whole of North America (542 million) and nearly as much as the whole of Europe (738 million).

According to OECD Economic Policy Paper No. 6 [1], in 2010, 60 % of global healthcare expenditures were directed towards people below 65 years old. In 2060, roughly the same percentage of expenditures will be directed to people aged above 65, reflecting an increase from 15 % to 30 % of their share in the total population.

As the proportion of the population at retirement age and beyond increases, the proportion in work is simultaneously decreasing, reducing the tax and national insurance base that supports healthcare services and further compounding the problem of supporting the increasing costs of treating and caring for the elderly. It is also important to note that, as people live longer, they have an increasing and justifiable expectation also to be able to maintain their dignity, independence and quality of life.

Will nanomedicine be able, therefore, to play a role where clinical and care needs are increasing due to this demographic shift and whilst health and care systems are under enormous pressure and costs are increasingly constrained? The following paragraphs provide some examples of research that has been funded by the European Commission and which is aimed at using nanotechnology, sometimes coupled with other enabling technologies, to address the health needs of an ageing population.

The European FP7 project NAD (Nanoparticles for Therapy and Diagnosis of Alzheimer's Disease), which commenced in 2008 and conclude in August 2013 is currently evaluating dendrimer nanocomposites for imaging and therapy, nanoliposomes for therapeutic agent delivery and other functionalized nanoparticles for applications in Alzheimer's disease [6].

The FP7 project Development of Novel *Nano*technology Based *Dia*gnostic Systems for *R*heumatoid *A*rthritis and Osteoarthritis (NanoDiaRA), which commenced in 2010 and was due to conclude in January 2014, is developing nanoparticle-based imaging and blood and urine-based diagnostic tools, and biomarkers, for the early detection of osteo- and rheumatoid arthritis. The research may also offer insights into the development of controlled nanoscale drug release and will consider the social, ethical and legal aspects of applying nanotechnology for medical purposes [7].

The FP6 integrated project Lidwine, which concluded its work in August 2010, developed novel approaches, including nanotechnology-treated textiles, for treating decubitus (pressure) ulcers, a painful and serious and, in terms of treatment, very common and expensive condition affecting many elderly bed- or chair-bound patients [8].

Moore (2011) reports other examples of nanotechnology research geared towards conditions affecting the elderly including the development of multifunctional nanoparticles capable of delivering controlled-release therapeutic agents to the inner ear for the treatment of age-related hearing loss and balance problems, and the use of nanotechnology in novel devices such as retinal implants for potential use in serious eye conditions such as macular degeneration [9].

Nanotechnology may also play a role in promoting the efficiency of care of the elderly through networked monitoring and telecare solutions which can be often be interfaced with novel biosensors incorporating micro- and nanotechnology. A variety of sensors can be embedded in the home, e.g. to monitor energy usage, movement or falls, or can be worn by the elderly person to monitor their physiological condition and provide a continuous feedback regarding their well-being or state of health to a remote monitoring station. Such networked systems can be used to alert health services or carers to react where there is an urgent or identified need, thereby allowing limited resources to be targeted more effectively as well as contributing to the independence of the patient.

Rather than being seen merely as an added cost, nanotechnologies should perhaps instead be viewed as a means of enabling novel healthcare and social care solutions and reducing the burden of long-term and expensive treatment of chronic conditions associated with ageing, as well as contributing to the dignity and independence of elderly persons.

19.4 Disruptive Innovation?

Clayton Christiansen (1997) defined several distinct types of innovation as follows:

Sustaining innovation: an innovation that does not affect existing markets.

Evolutionary innovation: an innovation that improves a product in an existing market in ways that customers are expecting.

Revolutionary (radical) innovation: an innovation that is unexpected, but which does not affect existing markets.

Disruptive innovation: an innovation that creates a new market by applying a different set of values, and which ultimately (and unexpectedly) overtakes an existing market [10].

Nanotechnology has the potential to impact medical products and processes at each of these levels. In many cases, the effects will be incremental such as improving the coating on an orthopaedic implant to improve its performance or lifetime or reformulating the delivery system of a drug to provide gradual release of that drug over an extended period.

The use of nanotechnology in new generations of devices such as retinal implants [11] could be considered an example of a revolutionary innovation in that it might have the potential to address currently unmet clinical needs such providing at least a limited level of vision for patients with macular degeneration or retinitis pigmentosa.

However, nanomedicine also has the potential for disruptive innovation. One example is its potential major contribution to the emerging field of regenerative medicine, for example the implantation of a nanostructured material into the body that can stimulate the body into self-repair producing new tissue such as in the regeneration of a damaged peripheral nerve [12] or the production of autologous bone that can be used elsewhere in the body for reconstructive surgery [13]. This type of emerging application may help shape a new future paradigm of medical treatment that could replace conventional treatments and for which major changes in procedures and training could be envisioned. Likewise the coupling of diagnostic and "-omics" tests (genomic, proteomic, metabolomic) with therapies (a concept sometimes referred to as *theranostics* [14–17] could herald a new, highly personalized form of medicine where, for example, the selection of drugs is matched to the individual patient and their condition, potentially reducing the considerable costs of prescribing drugs to patients for whom they have limited efficacy.

Whether the innovation brought about by nanomedicine is incremental, revolutionary or disruptive, there remains the potential for better treatments and lower costs but it is nevertheless important to consider the potential impacts on medical practice and procedures. Furthermore, while a progression towards a more personalized form of medicine may be strongly welcomed by both patients and medical professionals, it may not necessarily match the current business models of the major pharmaceutical and medical technology companies.

19.5   Risks and Regulatory Compliance

Protecting patients from risk is a primary objective of all medical product regulations but how this is actually achieved can vary widely in practice. In Europe, the regulation of the placing on the market of medical technologies is addressed primarily at the European level. In the US, the Food and Drug Administration (FDA) is primarily responsible. In nearly all countries around the world there are responsible national agencies or government departments.

Long-established product legislation was often drafted in a quite prescriptive style with a form of wording such as "you must not do this", "you shall do that and in this specific way". Many so-called "old approach" European Directives were drafted in this technical style and, as such, were not always adapted very well to areas of rapid innovation as the detailed requirements contained within the legal texts themselves could not always be changed quickly as new technological developments emerged. As this was recognized, newer types of product legislation, such as European "new approach" Directives, were developed which tended to be based around broad safety- and performance-based "essential requirements" rather than detailed prescriptive text, with the technical aspects being addressed in accompanying "harmonized" European standards drafted to support the broad essential requirements of the Directives. Such standards are, in theory, easier to revise if required although this can still be a lengthy process.

The approach taken by the various international agencies responsible for drug and device regulation varies. In the United States, the Food and Drug Administration (FDA) is responsible for determining the *primary mode of action* of the product and this decision will determine the regulatory framework for the product, i.e. a drug, medical device or biological product. The product regulatory application is thereafter managed by the appropriate FDA Center (Center for Drug Evaluation and Research – CDER; Center for Devices and Radiological Health – CDRH; Center for Biologics Evaluation and Research – CBER) with consultations from the other Centers.

In Europe, the *primary mode of action* of the product also determines the regulatory path(s) that will apply. Because European Directives are transposed into national legislation, national agencies and government departments have a responsibility for compliance within their jurisdiction.

The differences between what constitutes a medicinal product and what constitutes a medical device are similar in the US and Europe. In the US, products that have a primarily chemical/metabolic mode of action within the body are defined as drugs and, in Europe, products that have pharmacological, immunological or metabolic primary mode of action are defined as drugs and fall under the Medicinal Products Directive (2001/83/EC) or its related sister Directives or Regulations such as the Advanced Therapy Medicinal Products Regulation (Regulation EC No. 1394/2007). Similarly, in both regions, products that achieve their primary mode of action through physical or mechanical means are defined as medical devices and fall under their own regulatory pathways (the Medical Device Directives in the case of Europe). The European definition of a medical device (Article 1.2(a) of Directive 93/42/EEC) is as follows:

> ... *any instrument, apparatus, appliance, software, material or other article, whether used alone or in combination, together with any accessories... intended by its manufacturer to be used specifically for......*
>
> – *diagnosis, prevention, monitoring, treatment or alleviation of disease,*
> – *diagnosis, monitoring, treatment, alleviation of or compensation for an injury or handicap,*

- *investigation, replacement or modification of the anatomy or of a physiological process,*
- *control of conception,*

and which does not achieve its principal intended action in or on the human body by pharmacological, immunological or metabolic means, but which may be assisted in its function by such means. [18]

The FDA predicts that many nanotechnology-based products will span the regulatory boundaries between pharmaceuticals, medical devices and biologicals. It has stated [19] that technical assessments will be product-specific, taking into account the effects of nanomaterials in the particular biological and mechanical context of each product and its intended use, and that the particular policies for each product area, both substantive and procedural, will vary according to the statutory authorities. It also advises manufacturers to consult with the FDA early in their development process to facilitate a mutual understanding of the scientific and regulatory issues for their nanotechnology products.

With these points in mind, the FDA has issued guidelines describing its current thinking concerning regulated products containing nanomaterials or otherwise involving the application of nanotechnology. This guidance states that, based on its current scientific and technical understanding of nanomaterials and their characteristics, the FDA believes that evaluations of safety, effectiveness or public health impact of such products should consider the unique properties and behaviors that nanomaterials may exhibit [20].

In Europe, similar provisions exist for addressing combination products that may fall under more than one regulatory pathway. Since the primary mode of action may sometimes be difficult to determine for materials that exert an effect by virtue of novel properties arising at the nanoscale, determining the appropriate regulatory pathway(s) at an early stage of product development is of key importance.

In Europe, neither the Medicinal Products Directive nor the three Medical Device Directives (addressing medical devices, active implantable medical devices and in-vitro diagnostic medical devices, respectively) were originally drafted with nanotechnology in mind. The Medicinal Products Directive currently has no specific provisions relating to nanotechnology although a number of drugs containing nanomaterials have already been approved onto the European market. However, specific guidance on quality, toxicology, clinical development and monitoring aspects that have a bearing on nanotechnology are planned. Those developing drugs based on nanotechnology are strongly encouraged to interact with the relevant European Agency, the European Medicines Agency based in London which has an Innovation Task Force that addresses nanomedicine, from the earliest stages of development.

The European Medical Device Directives are based on broad "essential requirements" and the European Commission's Medical Devices Experts' Group has concluded that the provisions of the Directives broadly address nanotechnology-based medical devices. Essential requirements (ERs) of the Medical Device Directive [18] that are of general relevance to any technology and which can therefore apply equally to products based on nanotechnologies include the following:

> ER 1: *The devices must be designed and manufactured in such a way that, when used under the conditions and for the purposes intended, they will not compromise the clinical condition or the safety of patients, or the safety and health of users or, where applicable, other persons, provided that any risks which may be associated with their use constitute acceptable risks when weighed against the benefits to the patient and are compatible with a high level of protection of health and safety.*

> ER 2: *The solutions adopted by the manufacturer for the design and construction of the devices must conform to safety principles, taking account of the generally acknowledged state of the art. In selecting the most appropriate solutions, the manufacturer must apply the following principles in the following order:*
>
> - *eliminate or reduce risks as far as possible (inherently safe design and construction),*
> - *where appropriate take adequate protection measures including alarms if necessary, in relation to risks that cannot be eliminated,*
> - *inform users of the residual risks due to any shortcomings of the protection measures adopted.*

> ER 3: *The devices must achieve the performances intended by the manufacturer and be designed, manufactured and packaged in such a way that they are suitable for one or more of the functions.... as specified by the manufacturer. Any undesirable side-effect must constitute an acceptable risk when weighed against the performances intended.*

Other essential requirements address aspects such as chemical, physical and biological properties; infection and microbial contamination; construction and environmental properties; devices with a measuring function; protection against radiation; devices with an energy source; and accompanying information.

Two key themes in essential requirements 1–3 are those of *acceptable risk* and the *reduction of risk*. As there are broad knowledge gaps concerning the risks of many manufactured nanomaterials and, in many cases, a poor understanding of their novel properties and mechanisms of interaction with the body, this subject becomes an extremely important one in terms of compiling relevant information for regulatory approval, and the active collection or generation of appropriate data concerning risk and safety an essential activity in developing nanomedical products.

The European Commission's Scientific Committee on Emerging and Newly Identified Health Risks (SCENIHR) concluded in 2009 that a key limitation in the risk assessment of nanomaterials was the general lack of high quality exposure data both for humans and the environment. They noted that risk assessment procedures for the evaluation of potential risks of nanomaterials were still under development and could be expected to remain so until there is sufficient scientific information available to characterise the possible harmful effects on humans and the environment. They concluded that methodologies for both exposure estimations and hazard identification need to be further developed, validated and standardised [21].

A range of nanomaterial characteristics can give rise to novel hazards and their associated risks and these include particle size, shape, surface area, surface charge, surface chemistry, catalytic properties, solubility, crystalline phase, composition, zeta potential and other parameters. A useful overview of the issues surrounding the risk assessment of manufactured nanomaterials is given in the Organisation for Economic Co-operation and Development's 2012 Report *Series on the Safety of Manufactured Nanomaterials No. 33* [22]. Furthermore, international standards are currently in preparation that are intended to address some of these needs, such as those in ISO/TC 229 *Nanotechnologies* [23].

For medical devices, the harmonised standard EN ISO 14971 [24] describes a systematic risk management process that can be used as the basis for identifying hazards; analysing, estimating and reducing risks; deciding on the acceptability of risks; providing for post-manufacturing risk review; risk communication and risk documentation. While not specifically addressing nanotechnology or nanomedicine, with the addition of data on hazards and risks arising from the nanoscale characteristics of materials, it may nevertheless provide a useful basis for addressing risks for many medical devices incorporating nanotechnology .

One particularly important conclusion of this brief regulatory review is that there are still data gaps concerning the safety of many manufactured nanomaterials and, in the case of highly-regulated product sectors such as nanomedicine, that there is an urgent need to characterise nanomaterials and identify novel hazards and risks that arise from their nanoscale properties. In many cases this may also have implications for the development of new measurement and test methods, particularly those that can contribute towards characterising the interactions between nanoscale materials and biological systems *in-vivo* for nanomaterials that may come into contact with cells and tissues. This will form an important part of compiling risk data that will be required for subsequent regulatory approval.

19.6   Health Technology Assessment

Health technology assessment (HTA) has been defined as "a multi-disciplinary field of policy analysis that examines the medical, economic, social and ethical implications of the incremental value, diffusion and use of a medical technology in healthcare" [25]. Health technology assessment works together with, and relies on, many scientific disciplines such as epidemiology, biomedical sciences, behavioural sciences, clinical effectiveness studies, health economics, implementation science, health impact analysis and evaluation. As in the case of reimbursement systems, HTA systems vary from country to country.

Health technology assessment is intended to provide a bridge between research and decision-making. It is a growing field and is intended to provide the data to support management, clinical, and policy decisions. It is also underpinned by the development of various disciplines in the social and applied sciences, especially clinical epidemiology and healthcare economics. Health policy decisions are increasingly seen as important as the risk of incurring substantial costs from making wrong decisions grows with the rising costs of providing treatment. Evidence-based data and cost-effectiveness information from HTA is therefore increasingly-used in many countries to underpin such decision-making.

In 2004, the European Commission and Council of Ministers identified Health Technology Assessment as a political priority and decided that there was an urgent need to establish a sustainable European HTA network. A European network, EUnetHTA, was established to "...help develop reliable, timely, transparent and transferable information to contribute to HTAs in European countries". EUnetHTA comprises government-appointed organisations from the EU Member States, EEA and Accession countries, together with various regional agencies and non-for-profit organisations that produce or contribute to HTA in Europe [26].

At the global level, the International Network of Agencies for Health Technology Assessment (INAHTA) was established in 1993 and has now grown to include 57 member agencies from 32 countries including North and Latin America, Europe, Africa, Asia, Australia, and New Zealand. All its members are non-profit making organizations producing HTA and are linked to regional or national government. At a national level, most countries have a range of organisations dedicated to developing and implementing HTA methodologies. Notable examples of such bodies include the National Institute for Clinical Excellence (NICE) in the UK, the Institute for Quality and Efficiency in Healthcare (IQWIG) in Germany and the Agency for Healthcare Research and Quality (AHRQ) in the US.

Griffin [27] suggests that access to many European markets, following regulatory approval of a healthcare product, is controlled or influenced by HTA agencies whose decisions depend heavily on value arguments informed by evidence on relative benefits compared with existing standards of care, and by economic modelling. While the regulatory decision to approve a product onto the market or not is based on a scientific judgement of its risks and benefits, the HTA decision, which often also influences whether a technology will be reimbursed or not, is a value judgement, although one based on scientific evidence and economic data. This has broad implications for medical technology companies, whether in the pharmaceutical, device or diagnostic sectors. The intention of healthcare services (e.g. the UK National Health Service) that use HTA is for it to help contribute towards the most effective use of limited resources.

Following a survey of stakeholders, Stephens et al. [28] found that the most common type of cost analysis in HTA is cost-effectiveness, with the primary methodology being decision models. Common end points included cost/life-years saved, cost/event avoided and cost/quality-adjusted life years (QALY). European HTA agencies generally have defined national guidelines they follow, while US agencies are less consistent in this respect.

The same report goes on to conclude that the use of different research methods and their conformity to published HTA principles varies significantly from country to country. Despite the study's relatively small sample size, the results suggest that HTA, using evidence-based medicine, will continue to rapidly evolve and will need standardized research methods and principles to guide assessment and decision-making around novel drug therapies, medical devices, and emerging technologies. It suggests also that a process for information sharing among HTA bodies may be needed to achieve this standardisation in research methods.

The quality-adjusted life year (QALY), as used by NICE in the UK [29], is a measure of disease burden, including both the quality and the quantity of life lived. It is used in assessing the value-for-money of a medical intervention under consideration. The QALY is based on the number of years of life that would be added by the intervention. Each year in perfect health is assigned the value of 1.0 down to a value of 0.0 for death. If the extra years would not be lived in full health, then the extra life-years are given a value between 0 and 1 to account for this.

The measure is then used in a cost-utility analysis to calculate the ratio of cost to QALYs saved for a particular health care intervention. This is then used to allocate healthcare resources, with an intervention with a lower cost to QALY saved ratio being preferred over an intervention with a higher ratio.

The measure is not universally accepted – some opponents suggesting that it means that some people will not receive treatment where it is calculated that the cost is not warranted by the benefit to their quality of life. However, its supporters argue that since healthcare resources are inevitably limited, the measure enables them to be allocated in the way that is most beneficial to society rather than to an individual patient.

This review makes no value judgement about the use of QALYs or other HTA methodologies. Rather, attention is drawn to the increasing application of health technology assessment around the world as a process used to justify expenditure on novel medical technologies, and one that will certainly be applied to the emerging field of nanomedicine. There is, therefore, a clear need for companies to generate data during product development that can contribute towards this process.

19.7   Reimbursement and Novel Medical Technologies

In the development of any new medical technology, attention needs to be paid at an early stage to how that product will be taken up and paid for by healthcare systems and providers. In Europe, the reimbursement and pricing of medical products is determined on a country-by-country, rather than European-wide, basis, leading to significant variations in systems, costs, and availabilities.

Many developments in nanomedicine may facilitate progress towards personalizing treatment towards individual patients. In a review of the reimbursement of personalized medicine products in Europe, on behalf of the Personalized Medicine Coalition, Garfield (2011) found significant differences in the ability of different country's reimbursement infrastructures to effectively assess and provide access to novel personalized medicine technologies [30]. The report suggested that, as a result, healthcare systems in many countries have been failing to appropriately evaluate and pay for personalized medicine technologies, with patients often being denied access to the most advanced drug and diagnostic treatments, while those healthcare systems continue to bear the costs of outdated trial-and-error approaches to medicine.

Inbar (2012) suggests that the clinical data required for regulatory approval does not necessarily encompass the clinical data required for successful reimbursement of a medical product and there are large differences also in terms of cost and effort between fitting into an existing reimbursement mechanism and developing a new code. He states, however, that, in many cases, the data required for the reimbursement process can be developed in parallel to the required regulatory data during the same clinical trials and that companies that consider regulatory and reimbursement as serial processes may reach the market with insufficient funds and time to finance another clinical trial just to develop reimbursement-related data. He concludes that reimbursement needs to be viewed as one of the issues that needs to be dealt with in parallel and early in the device development process, adding that some mistakes may be very difficult and expensive to correct later on [31].

19.8 Professional Uptake of Nanomedicine

At the 2008 conference *The Future Delivery of Medicine: 2020*, hosted at University College London (UCL), one key finding was that the potential benefits of a range of new medical technologies were being delayed by slow uptake in many European national healthcare systems. It was noted that healthcare budgets were under pressure across Europe while, at the same time, new developments in science and technology have emerged that could transform medicine. It was further suggested that delivering this potential in an affordable way will require healthcare to be more patient-centered and for medical professionals to think beyond their specialities and take a far more holistic view [32].

In addition, at a meeting before the start of the main conference, a group comprising speakers and other experts discussed potential guidelines for future policy formulation, including

- a need for changes across the value chain, from basic research through to delivery of medical care at the bedside and in the home;
- fundamental rethinking and reshaping of all the processes that currently underpin healthcare systems;
- challenging healthcare professionals to look outside their specialities;
- requiring regulators to rethink their views of risk and reimbursement authorities to take a different view of value and affordability.

The participants at the meeting also suggested that that there was a need for a new view of value and noted that, while advanced treatments may be expensive, they can lead to cost savings elsewhere and that health technology assessments need to take a broader view in the face of this new paradigm.

While nanotechnology, as an enabling technology, and a continually-evolving understanding of how nanomaterials and biology interact at the nanoscale is beginning to revolutionise medicine and medical products in areas such as screening, prognosis, diagnosis, treatment planning, therapy, follow-up, and translational research, there is at present limited training available on nanomedicine, both within the curricula at medical schools and at a professional level thereafter.

A 2010 proposal to the European Commission's Directorate-General (DG) Research and Innovation Health Directorate by the European Alliance for Medical and Biological Engineering and Science (EAMBES) [33] suggested that the medical world could potentially become confused by the breadth and depth of the possible emerging medical technology interventions available. As a result, non-suitable solutions could be adopted that do not have the expected impact and thus do not constitute the correct way to approach the issue of preparing a framework for innovative therapeutic approaches. It suggested that this situation had already caused a number of problems in relation to the actual uptake of medical technology research and products resulting in a lower than expected synergy between the biological and medical engineering (BME) industry and the health sector.

The proposal went on further to suggest that a major impeding factor in the adoption of novel medical technology products is that they imply changes not only on the way the doctor thinks but also changes in the medical organizational and regulatory frameworks.

Many novel medical technologies have the potential to change the way medical practice is organised. Currently, typical diagnostic tests conducted by a General Practitioner might comprise taking a blood or bodily fluid sample from the patient, labelling and packaging it, sending it away to a central laboratory facility, waiting for several days for the results to come back and then recalling the patient to the surgery for a further consultation, discussion of the results of the tests and treatment. This multi-step procedure could potentially be replaced in the future by the use of a "smart" diagnostic device, designed for application in a variety of disease or metabolic tests, based on nanobiosensor and microfluidic technologies, and capable of being used in a GP's consulting room and of giving accurate results in a couple of minutes. Such novel diagnostic devices are currently in development and would, in all probability, be welcomed by GPs but there are a number of potential implications such as:

– diagnosis is changed from a remote dedicated laboratory facility/expert to a local "smart" device/medical generalist;
– while there may be costs in implementing such a technology, costs elsewhere, such as handling/packing/transport and laboratory costs would be minimized;
– long-established and familiar procedures would be changed;
– as diagnostic results could be immediately available, there would be implications for both GP, perhaps in terms of training on the interpretation of data and subsequent actions, and for patient;
– issues of trust in the quality and reliability of diagnostic data.

Therefore, in addition to the technological development of the device itself, attention needs to be paid to a broad spectrum of issues such as: the way and situation in which it will be used, e.g. by a patient at home, by a field worker or paramedic, by a qualified nurse, at a generalist's surgery or by a specialist at a hospital; whether existing practice or organisational aspects are altered; what implications this has for training, interpretation of results and consequent actions; impacts on costs and cost points; storage of confidential data; and many other aspects.

It is important, therefore, for researchers and companies to work with medical professionals at an early stage of product development. Nanomedicine, in particular, has implications for implementation by medical professionals as it utilizes properties of materials that manifest at the nanoscale and which may not be readily apparent or understood, or addressed in their training. Furthermore this understanding of the principles of nanomedicine by medical professionals is important as they form a key and trusted route of communication to patients.

19.9   Public Perception

Usually, the general public, as patients, will first come into contact with nanomedical products via medical professionals, with whom there is generally a high degree of trust and which, again, reinforces the importance of building relationships and trust with the medical profession during development of the product, as previously discussed.

The public's own perception of emerging technologies may be, however, influenced by previous scientific debates or controversies, such as "Mad Cow" Disease (bovine spongiform encephalopathy) (nvCJD), GMO foods, contaminated blood, etc., and how these have been represented, or misrepresented, in the popular media. The public cannot be expected to fully perceive and understand scientific risks arising from new technologies. The same public, however, are perfectly happy to take a risk/benefit decision where they broadly understand the factors involved and perceive the expected benefit as outweighing the risk, e.g. crossing the road, driving a car or travelling by air, or to choose one risk over another ("the lesser of two evils"). Many medical treatments are known by the public, as patients, to involve some measure of risk, e.g. X-rays or aggressive chemotherapy, but they are prepared to undergo such procedures as they perceive the benefits to be gained as outweighing those risks and trust those professionals that carry out such procedures.

The perception of a risk amongst the general public can vary greatly depending upon factors such as:

- the cultural, socio-economic and educational background of the person(s) involved
- whether exposure to the hazard is

  - involuntary;
  - avoidable;
  - from a man-made or natural source;
  - due to negligence;
  - arising from a poorly understood cause;
  - affecting a vulnerable group within society;

- whether there is an obvious benefit to be gained from exposure to the risk.

Furthermore there may be a tendency to distrust "big industry" in some sectors where profits may be seen to outweigh safety concerns. All of these factors taken together may colour attitudes towards the acceptance of new technologies, especially if there has been poor communication about them.

Kahan and co-workers (2007) carried out a study amongst a recruited sample of United States subjects to assess their opinions about nanotechnology [34]. The responses of 1,500 subjects not exposed to additional information suggested that Americans were largely uninformed about nanotechnology: 81 % of subjects reported having heard either "nothing at all" (53 %) or "just a little" (28 %) about nanotechnology prior to being surveyed, and only 5 % reported having heard "a lot." Nevertheless, most of the same group of subjects, 89 %, were reported as having an opinion on whether the benefits of nanotechnology outweigh its risks or vice versa with slight majority (53 %) appearing to view benefits as outweighing risks. When subgroups were examined, however, more divisions were revealed. Men (59 % to 36 %) were significantly more likely than women (47 % to 40 %) to think that risks outweigh benefits. Moreover, whereas a majority of whites (54 %) believed that benefits outweighed risks, 49 % of African-Americans of viewed risks as outweighing benefits. White males were the most pro-benefit orientated (61 % to 30 %).

The study also backed up conclusions from previous studies that *affect* (a person's positive or negative emotional orientation) is one of the most powerful influences on individuals' perceptions of risk – subjects in the survey were asked to indicate whether nanotechnology made them feel "very bad," "bad," "neither good nor bad," "good," or "very good." Furthermore the study suggested how people react to information depends largely on their *values*. One of the major findings was that dissemination of scientifically-sound information is not by itself sufficient to overcome the divisive tendencies of cultural cognition. The authors concluded that those in a position to educate the public, including government, scientists and industry, need also to intelligently frame that information in ways that make it possible for persons of diverse cultural orientation to reconcile it with their values.

A later study by Bottini and colleagues (2011) amongst 790 citizens chosen randomly from four different urban areas of Rome reported that those surveyed exhibited optimism towards nanomedicine despite low awareness of currently available nanodrugs and nanocosmetics, and limited understanding of biocompatibility and toxicity aspects. The study concluded that, if such public optimism justifies the increase in scientific effort and funding for nanomedicine, it also obliges toxicologists, politicians, journalists, entrepreneurs, and policymakers to be more responsible in their dialogue with the public [35].

While there would seem, therefore, to be no major widespread prior distrust of the application of nanotechnology to medicine despite concerns in other areas of technology there is, nevertheless, a need for clear information to be made available to the public and other stakeholders about the benefits and risks of nanomedicine in a language that can be clearly understood and through channels that are trusted.

19.10   Ethical Considerations and Safeguards

While an in-depth review of many of the potential ethical issues associated with nanomedicine is provided by Donald Bruce within this book, it is nevertheless useful to consider some key points here as part of an overview of the challenges facing its widespread implementation.

19.10.1   What Do We Understand by Healthcare?

The increasing ability that we have to manipulate matter precisely at the nanoscale, combined with our improved understanding of biology, may influence our perception of what medicine and what a well person is, e.g.

– Just the treatment of disease?
– The correction of any deviation from what is considered "normal" function?
– What do we mean by "well" if we will be able to monitor at so many levels?
– What is the borderline between impaired function correction and performance enhancement?
– What are the expected limits of a "cure"?

While most people would probably accept the use of medicine for treatment of a disease or the correction of a physiological condition or impairment, they may not readily accept its application for enhanced performance, e.g. strength, senses, endurance for sports, military or other non-medical purposes.

19.10.2   The Changing Face of Medicine

Over the past several centuries medicine has changed beyond all recognition from the seventeenth century where treatments were largely palliative with the doctor focusing mainly on nonphysical supportive measures, through the development of hospital medicine in the nineteenth century and "laboratory medicine" in the twentieth century to the current twenty-first century scenario where we are now beginning to understand the human body as an intricately structured machine with billions of complex interacting parts, with each part (and each subsystem of parts) potentially subject to individual investigation, repair, and possibly replacement by artificial technological means. Along with this transformation of medicine over the centuries, the role of the medical professional has also changed enormously and we might reasonably expect medicine to become even more technological. But do good scientists or engineers make good doctors?

19.10.3   A Data Overload?

The development of novel diagnostic and imaging technologies, coupled with advances in genomics, proteomics and metabolomics (now commonly referred to collectively as the "-omics") means that there is a huge amount of data becoming available to medical professionals. This begs important questions such as

- Who can interpret all of this data?
- How much of the information is clinically significant?
- Who does the data belong to? The healthcare provider? The patient?
- How will the data be stored and transmitted safely?
- Where will the data be stored?
- What about patient confidentiality issues?
- What about the patients right to *know* and, equally, *not to know* certain information?

One particularly important element is maintaining the confidentiality of medical data… much of it could be of value to third parties other than the patient and doctor, e.g. employers, the government and commercial organisations such as insurance companies.

A study by Erlich and colleagues (2012) at the USA's Whitehead Institute demonstrated that the supposedly confidential names of research study participants could be traced from de-identified genetic data [36]. The researchers identified nearly 50 men and women who had submitted samples and had their genomes sequenced for a study performed by the Center for the Study of Human Polymorphisms (CEPH).

By matching short tandem repeats that they found on the Y chromosomes of men in the CEPH study to Y-STRs in publicly-available genetic genealogy databases, the researchers were able to recover the family names of men in the CEPH dataset who had submitted their Y-STRS to these repositories. With this information, they searched other free online information sources including record search engines, obituaries, genealogy websites, and public demographic data from the National Institute of General Medical Sciences' Human Genetic Cell Repository, housed at the Coriell Institute, and were able to track down the participants.

This study suggests that it may be difficult in practice to guarantee the security of medical and genomic data and that there is a need to balance research participants' privacy rights with the societal benefits to be realized from the sharing of biomedical research data.

19.10.4   Non-discrimination and Equity

Non-discrimination is a widely-accepted principle that people deserve equal treatment unless there are reasons that justify difference in treatment. In this context it primarily relates to the distribution of healthcare resources. Equity is the ethical principle that everybody should have fair access to the benefits under consideration.

Earlier commentary indicates, however, that access to treatment may vary from country to country because of regulatory, health technology assessment and reimbursement issues and, within some countries, access may even vary between different regions due to differing practices, priorities or availability of resources.

19.10.5 The Precautionary Principle

This principle entails the moral duty of continuous risk assessment with regard to the not fully foreseeable impact of new technologies. While the Precautionary Principle is already enshrined within European legislation, there are concerns from some quarters that could it be used as the justification to block potentially life-saving technologies on the grounds that the science is not yet fully understood.

19.11 Training

In formal medical education, very few medical schools currently offer modules on nanomedicine as part of their curricula. At the same time, massive levels of investment into research on the application of nanotechnologies to medicine, at both academic and commercial levels, means that there are increasing numbers of products incorporating nanotechnology appearing on the market with many more in the product pipeline or at the stage of clinical trials or awaiting regulatory approval.

While an increasing number of universities are now offering nanotechnology-based undergraduate or postgraduate level courses, there are still only a limited number specifically addressing nanomedicine or specific medical disciplines with a significant medical nanotechnology component.

At a professional level, organisations such as such as the Institute of Nanotechnology (IoN) and universities such as Cranfield and Oxford have developed short courses aimed at addressing training needs in nanomedicine and the application of nanotechnology to topics such as medical diagnostics, imaging, drugs and biosensors, as well as nano- risk and safety issues. These have attracted interest from a range of participants including those working in academia and research, industry, medical professionals and medical students, healthcare providers and regulatory authorities.

The successful adoption and implementation of nanomedical solutions in the clinic will depend on the presence of informed decision-makers who understand the underlying science, opportunities and benefits that the technologies can bring, short and long terms costs and savings, and how nanomedicine can be integrated safely and effectively into everyday healthcare. This includes those working in research funding, commercial strategy, regulatory affairs, health technology assessment, reimbursement and healthcare provision professionals, insurers, and amongst the medical professions. There is, therefore, an ongoing need for training in nanomedicine at both academic and in-service, professional levels.

19.12   Conclusions and Perspectives for the Future

This chapter intends to highlight some of the non-technical challenges that researchers and developers are likely to face in bringing medical products based on nanotechnology to the market and clinic. Many of these challenges are not exclusive to nanomedicine but apply generally to emerging medical technologies. However, some of these challenges may be compounded by the fact that nanoscale materials frequently exhibit novel properties that can provide both benefits and opportunities but that, at the same time, may present novel hazards and risks that are poorly understood. Characterisation of novel nanomaterials and the establishment of a widely-available repository of safety data will therefore be vital to the success of nanomedicine.

From the author's personal experience, the attitude towards nanomedicine from a wide variety of stakeholders who have attended professional training courses, workshops and conferences on the topic, including medical professionals, regulators, industry professionals and others, has been positive. There, however, remains a widespread lack of awareness on the subject in the wider medical community and much needs to be done to engage with these professionals to impart knowledge, build trust and promote the uptake of novel nano-based products.

It is also clear that better communication is needed with health technology and reimbursement professionals. In healthcare systems where cost containment is increasingly critical to healthcare delivery, it must clearly be demonstrated that nanomedicine can deliver better treatments while reducing costs in the short, middle or long term, for example by earlier or more accurate diagnosis, more effective treatments, or by reducing lengths of stay in hospitals. In addition, there is clear scope for a contribution towards more personalised form of medicine rather than a one-size-fits-all approach, although this may well necessitate the development of new business and professional practice models.

Because of the comparative timescales required for regulatory approval, it is likely that the fastest progress to market for nanomedicine will be seen in the areas of diagnostics, biosensors and other medical devices. However, developments in the pharmaceutical and regenerative medicine sectors, although possibly longer term, are likely to be significant and potentially disruptive in terms of contributing to new paradigms of treatment.

In the longer term, there is also potential for massive synergy between nanomedicine and other emerging field such biomimetics, particularly in terms of integrating nano- and biological structures for biosensing, drug delivery and regenerative medicine, and designing new generations of novel nano-based devices.

References

1. De la Maisonneuve C, Oliveira Martins J (2013) Public spending on health and long-term care: a new set of projections, OECD Economic policy paper no. 6. http://www.oecd.org/eco/growth/Health%20FINAL.pdf. Accessed 4 July 2013
2. Robinson JC, Smith MD (2008) Cost-reducing innovation in health care. Health Aff 27(5):1353–1356. http://content.healthaffairs.org/content/27/5/1353.full. Accessed 15 July 2013

3. Krishna Kumar R (2011) Technology and healthcare costs. Ann Pediatr Cardiol 4(1):84–86.. http://www.ncbi.nlm.nih.gov/pmc/articles/PMC3104544/. Accessed 15 July 2013

4. Communication from the Commission to the European Parliament, the Council, the European Economic and Social Committee and the Committee of the Regions, Dealing with the impact of an ageing population in the EU (2009 Ageing Report), Commission of the European Communities, Brussels, 29 Apr 2009. COM (2009) 180 final. http://ec.europa.eu/economy_finance/publications/publication13782_en.pdf. Accessed 20 July 2013

5. World Alzheimer Report (2012) Overcoming the stigma of dementia. In: Batsch NL, Mittelman MS (eds) Alzheimer's disease international. http://www.alz.co.uk/research/WorldAlzheimer Report2012.pdf. Accessed 4 July 2013

6. Nanoparticles for Therapy and Diagnosis of Alzheimer Disease (2013) EU framework programme 7. http://www.nadproject.eu/. Accessed 17 July 2013

7. Development of Novel Nanotechnology Based Diagnostic Systems for Rheumatoid Arthritis and Osteoarthritis (2013) EU framework programme 7. http://www.nanodiara.eu/. Accessed 17 July 2013

8. Multi-functionalised medical textiles for wound (e.g. decubitus) prevention and improved wound healing (2010) EU framework programme 6. http://cordis.europa.eu/projects/rcn/81553_en.html. Accessed 17 July 2013

9. Moore R (2011) Ageing with confidence – the capacity of advanced technology to support the increasingly ageing population. Public Serv Rev Eur Union (22):550. http://www.publicservice.co.uk/article.asp?publication=European%20Union&id=525&content_name=Science,%20Research%20and%20Technology&article=17595. Accessed 17 July 2013

10. Christensen CM (1997) The innovator's dilemma: when new technologies cause great firms to fail. Harvard Business School Press, Boston. ISBN 978-0-87584-585-2

11. http://retina-implant.de/en/default.aspx. Accessed 22 July 2013

12. Zhan X, Gao M, Jiang Y, Zhang W, Wong WM, Yuan Q, Su H, Kang X, Dai X, Zhang W, Guo J, Wu W (2013) Nanofiber scaffolds facilitate functional regeneration of peripheral nerve injury. Nanomedicine 9(3):305–315. http://www.ncbi.nlm.nih.gov/pubmed/22960189. Accessed 22 July 2013

13. Stevens MM, Marini RP, Schaefer D, Aronson J, Langer R, Shastri VP (2005) In vivo engineering of organs: the bone bioreactor. Proc Natl Acad Sci USA 102:11450–11455. http://www.pnas.org/content/102/32/11450.abstract. Accessed 22 July 2013

14. Chen G, Qiu H, Prasad PN, Chen X (2014) Upconversion nanoparticles: design, nanochemistry, and applications in theranostics. Chem Rev 114(10):5161–5214

15. Liu Y, Yin JJ, Nie Z (2014) Harnessing the collective properties of nanoparticle ensembles for cancer theranostics. Nano Res. doi:10.1007/s12274-014-0541-9

16. Lu ZR (2014) Theranostics: fusion of therapeutics and diagnostics. Pharm Res 31(6):1355–1357

17. Muthu MS, Leong DT, Mei L, Feng SS (2014) Nanotheranostics application and further development of nanomedicine strategies for advanced theranostics. Theranostics 4(6):660–677

18. Council Directive 93/42/EEC of 14 June 1993 concerning medical devices, Official Journal of the European Communities, L 169, 12 June 1993, 0001–0043. http://eur-lex.europa.eu/LexUriServ/LexUriServ.do?uri=CELEX:31993L0042:EN:HTML. Accessed 22 July 2013

19. US Food and Drug Administration (2012) FDA's approach to regulation of nanotechnology products. http://www.fda.gov/ScienceResearch/SpecialTopics/Nanotechnology/ucm301114.htm. Accessed 22 July 2012

20. US Food and Drug Administration (2011) Considering whether an FDA-regulated product involves the application of nanotechnology – guidance for industry. http://www.fda.gov/RegulatoryInformation/Guidances/ucm257698.htm. Accessed 22 July 2013

21. Risk Assessment of Products of Nanotechnologies (2009) Scientific Committee on Emerging and Newly Identified Health Risks (SCENIHR), European Commission Directorate-General for Health and Consumers. http://ec.europa.eu/health/ph_risk/committees/04_scenihr/docs/scenihr_o_023.pdf. Accessed 17 July 2013

22. Important Issues on Risk Assessment of Manufactured Nanomaterials (2012) Series on the safety of manufactured nanomaterials no. 33. Organisation for Economic Co-operation and Development (OECD), Paris. http://search.oecd.org/officialdocuments/displaydocumentpdf/?cote=env/jm/mono(2012)8&doclanguage=en. Accessed 17 July 2013

23. ISO/TC 229 "Nanotechnologies" Work Programme (2013) International Standards Organization, Geneva. http://www.iso.org/iso/home/store/catalogue_tc/catalogue_tc_browse.htm?commid=381983&development=onAccessed. 17 July 2013

24. EN ISO 14971 (2012) Medical devices. Application of risk management to medical devices. http://shop.bsigroup.com/en/ProductDetail/?pid=000000000030268035. Accessed 20 July 2013

25. International Network of Agencies for Health Technology Assessment (INAHTA) website (2013) http://inahta.episerverhotell.net/HTA/. Accessed 14 July 2013

26. EUnetHTA website (2013) http://www.eunethta.eu/. Accessed 14 July 2013

27. Griffin A (2011) An industry perspective on HTA. Regul Rapp 8(4), TOPRA. http://www.topra.org/sites/default/files/regrapart/1/3521/focus1.pdf. Accessed 14 July 2013

28. Stephens JM, Handke B, Doshi JA (2012) International survey of methods used in health technology assessment (HTA): does practice meet the principles proposed for good research? Comp Eff Res 2012:2, 29–44. Dove Medical Press. http://www.ispor.org/workpaper/International-survey-of-methods-used-in-HTA.pdf. Accessed 15 July 2013

29. National Institute for Clinical Excellence (2010) Measuring effectiveness and cost effectiveness: the QALY. http://www.nice.org.uk/newsroom/features/measuringeffectivenes-sandcosteffectivenesstheqaly.jsp. Accessed 22 July 2013

30. Garfield S (2011) Advancing access to personalized medicine: a comparative assessment of european reimbursement systems. Personal Med Coalit. http://www.personalizedmedicinecoalition.org/sites/default/files/files/PMC_Europe_Reimbursement_Paper_Final.pdf. Accessed July 18 2013

31. Inbar A (2012) It's never too soon to start thinking about reimbursement. Eur Med Device Technol. http://www.emdt.co.uk/article/it%E2%80%99s-never-too-soon-start-thinking-about-reimbursement. Accessed 20 July 2013

32. UCL Media Relations (2008) Medical benefits being delayed by slow uptake of new medical technologies. http://www.ucl.ac.uk/media/library/Medtechnologies. Accessed 15 July 2013

33. Maglaveras N, Viceconti M (2010) Bridging technology with therapeutics for innovative therapeutics intervention – a proposal to the DGRTD HEALTH for consideration for future calls. EAMBES. http://www.eambes.org/contents/public-repository/DGRT_EAMBES_proposal_062010_final.pdf. Accessed 15 July 2013

34. Kahan DM, Slovic P, Braman D, Gastil J, Cohen G (2007) Nanotechnology risk perceptions: the influence of affect and values, cultural cognition project at Yale Law School/Project on Emerging Nanotechnologies at the Woodrow Wilson International Center for Scholars. http://www.nanotechproject.org/file_download/files/NanotechRiskPerceptions-DanKahan.pdf. Accessed 15 July 2013

35. Bottini M, Rosato N, Gloria F, Adanti S, Corradino N, Bergamaschi A, Magrini A (2011) Public optimism towards nanomedicine. Int J Nanomed 6:3473–3485. http://www.ncbi.nlm.nih.gov/pmc/articles/PMC3260040/. Accessed 22 July 2013

36. Gymrek M, McGuire AL, Golan D, Halperin E, Erlich Y (2012) Identifying personal genomes by surname inference. Science. Referred to in article from the Whitehead Institute available at http://wi.mit.edu/news/archive/2013/scientists-expose-new-vulnerabilities-security-personal-genetic-information. Accessed 22 July 2013

Biography of the Editors

Dr. Yi Ge obtained his bachelor's degree (first Class Hons) in Biopharmaceutics and went on to an MPhil degree in Pharmaceutical Chemistry at Aston University. Afterwards, he moved to the University of Sheffield for a PhD in Chemistry. He was later employed as a research scientist in a UK pharmaceutical company and was then a postdoctoral research associate at Imperial College London, before joining Cranfield University as a member of academic staff in 2006. His research interests and activities have a focus on the interdisciplinary areas, at the interfaces of nanotechnology, medicine, materials and pharmaceutics. In 2008, as the founder, he was appointed as the Course Director of Nanomedicine MSc Course Programme, which is a unique postgraduate course and the first of its kind in all of Europe. He is a Member of the Royal Society of Chemistry and a Member of the British Society for Nanomedicine. His activity in the field of nanotechnology and nanomedicine was recognized by the Institute of Nanotechnology and he was admitted as a Fellow of the Institute in 2009. He is a Visiting Professor at the University of Jinan (China) and serves as the Associate Editor of *Smart Materials*, as well as the Editorial Board Members of the *Journal of Nanotechnology: Nanomedicine and Nanobiotechnology*, *Austin Journal of Nanomedicine and Nanotechnology* and *Austin Journal of Biosensors and Bioelectronics*. He has also been the Editor of two books on Smart Nanomaterials for Sensor Applications (Bentham Science) and Molecularly Imprinted Sensors (Elsevier).

Dr. Songjun Li was appointed as an Associate Professor Central China Normal University after receiving his PhD degree from the Chinese Academic of Sciences in 2005. In 2008, he took up a position at the University of Wisconsin-Milwaukee (USA) as a Postdoctoral Associate until 2009 when he received a prestigious Marie Curie International Incoming Fellowship to undertake a 2-year research project at Cranfield University (UK). He is now a Distinguished Professor of Functional Polymers at Jiangsu University in China. Dr. Li's research interests and activities, in relation to nanomedicine, includes nanomaterials, nanobiosensors and smart nanoreactors. He has published over 60 papers in peer-reviewed journals and has edited

© Springer Science+Business Media New York 2014
Y. Ge et al. (eds.), *Nanomedicine*, Nanostructure Science and Technology,
DOI 10.1007/978-1-4614-2140-5

five books with Wiley-VCH, Elsevier, Bentham Science, Nova Science and Research Signpost. He also serves as the Editorial Board Members of several journals, such as *Chemical Sensors*, *American Journal of Environmental Sciences*, *Journal of Public Health and Epidemiology*, and *Journal of Computational Biology and Bioinformatics Research*.

Richard Moore has been active in the fields of medical technology and nanotechnology for over 20 years. His experience includes: (1) 6 years working as Scientific Director at the Institute of Nanotechnology, UK, with a particular focus on the application of nanotechnologies to medicine and the life sciences, risk management, the regulation and governance of nano- and other novel technologies and the development of professional training courses in nanomedicine; (2) 10 years working as Director, Science and Innovation at Eucomed (European Medical Technology Association), Brussels, with a key focus on providing technical and scientific support services to Eucomed's national and corporate members; (3) 6 years working as Project Manager, Healthcare at the European Committee for Standardisation (CEN), Brussels, with a main focus on overseeing the development of the platform of European harmonized standards supporting the Medical Device Directives. Since early 2012, Richard Moore has run his own consultancy specialising in analysis, foresighting, and advisory services in the field of emerging technologies including nanotechnology. He is also a Project Technical Advisor to a number of large EC-funded research projects in the areas of nanosafety and advanced nanomaterials, and is an external lecturer on nanotechnology and nanomedicine topics at several UK universities.

Dr. Shenqi Wang earned his PhD in Materials Science from the Biomedical Materials and Engineering Centre at Wuhan University of Technology (China) in 2000. He was then worked as a Postdoctoral Fellow and Research Associate Professor at the Key Laboratory of Bioactive Materials at Nankai University (China). In 2006, he moved to the Division of Bioengineering at Nanyang Technological University in Singapore as a Research Fellow for a research project on the optical fiber-based nanobiosensors. Dr. Wang was later appointed as a Full Professor at Huazhong University of Science and Technology (China) in 2008 and then awarded a prestigious Marie Curie International Incoming Fellowship to undertake a 2-year research project in the field of nanomedicine and nanobiosensor at Cranfield University (UK) in 2010. He is a Member of the World Biomaterials Society and a Member of the Optics Society (USA). He also holds Memeberships of the Chinese Biomedical Engineering Society, the Chinese Materials Society and the Chinese Biomedical Materials. His research interests and activities involve nanomaterials, biomaterials, nanobiosensors and bio-interactions with materials.

Index